To Doug

I'm sure this book will be interesting
it shows some of the least known
parts of the Chilean water law,
and also makes frequent reference
to how the Bank has used this law
to advance its agenda

Abel

June 2004

SIREN SONG

Chilean Water Law
as a Model for
International Reform

Carl J. Bauer

Resources for the Future
Washington, DC, USA

Copyright ©2004 by Resources for the Future. All rights reserved.
Printed in the United States of America

No part of this publication may be reproduced by any means, whether electronic or mechanical, without permission. Requests to photocopy items for classroom or other educational use should be sent to the Copyright Clearance Center, Inc., Suite 910, 222 Rosewood Drive, Danvers, MA 01923, USA (fax +1 978 646 8600; www.copyright.com). All other permissions requests should be sent directly to the publisher.

An RFF Press book
Published by Resources for the Future
1616 P Street, NW
Washington, DC 20036–1400
USA
www.rffpress.org

Library of Congress Cataloging-in-Publication Data

Bauer, Carl J., 1960–
Siren song : Chilean water law as a model for international reform / Carl J. Bauer.
p. cm.
ISBN 1-891853-79-1 (hardcover : alk. paper)
1. Water—Law and legislation—Chile. 2. Water—Government policy—Chile.
3. Water-supply—Chile. 4. Water—Government policy. I. Title.
KHF3310.B38 2004
346.8304'691—dc21 2004000928

f e d c b a

The paper in this book meets the guidelines for permanence and durability of the Committee on Production Guidelines for Book Longevity of the Council on Library Resources.

This book was typeset in Giovanni and Myriad by Betsy Kulamer. Interior design by Naylor Design Inc. Permission to modify the maps, which originally appeared in *Against the Current: Privatization, Water Markets, and the State in Chile,* by Carl J. Bauer, was kindly granted by Kluwer Academic Publishers.

The findings, interpretations, and conclusions offered in this publication are those of the author and are not necessarily those of Resources for the Future, its directors, or its officers.

The geographical boundaries and titles depicted in this publication, whether in maps, other illustrations, or text, do not imply any judgment or opinion about the legal status of a territory on the part of Resources for the Future.

ISBN 1-891853-79-1 (cloth)

About Resources for the Future *and* RFF Press

Resources for the Future (RFF) improves environmental and natural resource policymaking worldwide through independent social science research of the highest caliber. Founded in 1952, RFF pioneered the application of economics as a tool to develop more effective policy about the use and conservation of natural resources. Its scholars continue to employ social science methods to analyze critical issues concerning pollution control, energy policy, land and water use, hazardous waste, climate change, biodiversity, and the environmental challenges of developing countries.

RFF Press supports the mission of RFF by publishing book-length works that present a broad range of approaches to the study of natural resources and the environment. Its authors and editors include RFF staff, researchers from the larger academic and policy communities, and journalists. Audiences for publications by RFF Press include all of the participants in the policymaking process—scholars, the media, advocacy groups, NGOs, professionals in business and government, and the public.

RESOURCES FOR THE FUTURE

Directors

Catherine G. Abbott
Joan Z. Bernstein
Julia Carabias Lillo
Norman L. Christensen Jr.
Maureen L. Cropper
W. Bowman Cutter
John M. Deutch
E. Linn Draper Jr.
Mohammed El-Ashry
Dod A. Fraser
Kathryn S. Fuller
Mary A. Gade
David G. Hawkins
Lawrence H. Linden
Lawrence U. Luchini
Jim Maddy
James F. O'Grady Jr.
Steven W. Percy
Mark A. Pisano
Robert N. Stavins
Joseph E. Stiglitz
Edward L. Strohbehn Jr.

Officers

Robert E. Grady, Chair
Frank E. Loy, Vice Chair
Paul R. Portney, President
Edward F. Hand, Vice President, Finance and Administration
Lesli A. Creedon, Vice President, External Affairs and Corporate Secretary

---

***Editorial Advisors for* RFF Press**

Walter A. Rosenbaum, University of Florida
Jeffrey K. Stine, Smithsonian Institution

Map 1. Central Chile

CONTENTS

List of Maps

PREFACE

This is my second book about water law and policy in Chile. It is the result of my research and work experience from the mid 1990s to early 2004, particularly in Chile and the United States, but also in several other Latin American countries and in Spain.

Siren Song differs from my previous book, *Against the Current,* in covering a longer period and taking a more comparative and international perspective. *Against the Current* was an in-depth case study based on two and a half years of fieldwork in Chile (during two periods from 1991 to 1995). *Siren Song* builds on that foundation by continuing to track water rights issues in Chile since 1995 and by placing the Chilean experience more fully in the international context. This book, in other words, has a broader focus and argument and is intended for a broader readership.

I have several main purposes in this book. In the first place, I aim to present an up-to-date overview of Chile's experience with what has become known internationally as the Chilean model of water resources management. This is a historical overview in the sense that it covers several recent decades and focuses on evolution over time. In the second place, I aim to highlight the significance of the Chilean model for international debates about water policy. These debates have become more urgent in recent years as people worry more about a "global water crisis"—a crisis that seems doomed to get worse unless we make major changes in how we use and manage the world's water. In that context the Chilean model has been like a song of the sirens to would-be reformers in other countries, so alluring in its free-market purity that many have been deaf to the risks below the surface.

The structure of this book reflects my own personal and professional trajectory over the past 10 to 15 years. Since the early 1990s I have worked intensively in Chile while also steadily gaining experience in other countries and parts of the world. As a result, I have wanted to tell the story of contemporary Chilean water policy both from the inside, for outsiders, and from the outside, for insiders. One part of the story grows out of my longstanding efforts to understand the local complexities of Chilean politics, culture, and geography with as much historical richness and messy factual detail as I can manage. I began these efforts in earnest in 1991 and I have continued to the present day. On the other hand, my analysis of Chilean water issues has been ever more influenced by my growing knowledge of the water problems faced by other countries and regions. As anyone who has done comparative work will understand, the problems in other places are similar in many ways but are shaped by different political, economic, geographic, and historical circumstances.

In this book, therefore, I have wanted to counter the superficiality of most foreign descriptions of Chilean water law and policy by doing justice to the deeper, tangled political roots that underlie what may seem to be merely technocratic discussions. At the same time, I have wanted to show Chileans who are enmeshed in local struggles and challenges what those issues look like from a distance, to an outside observer familiar with water issues in other places. From both perspectives, internal and external, I hope that readers find the story useful and illuminating and that some of its drama comes through in the telling.

My third purpose is more general: to make a pitch for a more qualitative and interdisciplinary approach to law and economics. I do not offer an elaborate theoretical argument for this. Instead, I try to use water as a vehicle to highlight the shortcomings of traditional academic disciplines when applied to the wider world. Since the larger goal, in my view, is to try to understand the world and the human condition as fully as possible, we must keep in mind that individual disciplines are only means to that end.

ACKNOWLEDGMENTS

I am grateful to my employers at Resources for the Future for their support of my work since 1999, both in Washington, DC, and overseas. The flexibility and encouragement of Terry Davies, Ted Hand, Ray Kopp, Paul Portney, and Mike Taylor (in alphabetical order) have been essential to my carrying out the research that led to this book and to related publications and forms of public communication. I am also grateful to the William and Flora Hewlett Foundation for two grants during 2000–2003, in support of my research and collaboration with local universities and nongovernmental organizations in Chile and elsewhere in Latin America. Without the support of David Lorey, former director of the foundation's U.S.–Latin American Relations Program, this book would not have been written.

In Chile I am grateful to former colleagues at the United Nations Economic Commission for Latin America and the Caribbean, in Santiago, where I have been a regular visitor and occasional consultant over the years; Martine Dirven, Terry Lee, Miguel Solanes, Pedro Tejo, and Frank Vogelgesang have always been helpful. I wrote much of this book in Santiago in 2002 while I was a visiting scholar at the Center of Applied Economics at the University of Chile; I want to thank Ronnie Fischer, the Center's director, and his colleagues Alex Galetovic, Patricio Meller, and Raúl O'Ryan for their hospitality during my stay.

I am also grateful to the many people in Chile who agreed to be interviewed and otherwise facilitated my work and my access to information over the past 12 years. Although these people are too numerous to name individually, there are two friends whose generosity and cooperation deserve special mention: Alejandro Vergara Blanco at the Catholic University and Luis Catalán Torres at the University of Chile. Vergara's name appears frequently in this book, since he is Chile's foremost scholar of water law as well as a leading practitioner, and although our views are often different, his collegiality has never wavered. Catalán has provided me with a customized clipping service about so many issues in Chile and for so long that I have often had trouble keeping up with him.

During 2001–2003 I presented earlier versions of the arguments in this book at seminars and conferences in Chile and the United States. In Santiago these fora included the University of Chile Center of Applied Economics; the Catholic University Law School (4th and 5th Chilean Water Law Conferences); the Diego Portales University Law School (7th Conference of Latin American Law and Economics Association); and the Chilean government's Institute for Agricultural Development. U.S. fora included seminars at Resources for the Future and the Inter-American Development Bank, in Washington, DC; and the University of Colorado Law School in Boulder (23rd Summer Water Law Conference). I want to thank the participants at those events for their comments.

At RFF Press I appreciate the able work of Don Reisman, Jessica Palmer, and Carol Rosen. Ken Conca (University of Maryland) and Chuck Howe (University of Colorado) provided comments that significantly improved the manuscript, and Sally Atwater's copyediting improved it further. Former RFF Research Assistant Aracely Alicea helped with references. All translations from Spanish are my own.

I would like to dedicate this book to Brooke, my wife and manager, and to Hugo and Halle Mar, who were both born in Chile and shared different stages of the writing.

INTRODUCTION: THE CHILEAN WATER MODEL COMES OF AGE

Fresh water is a finite and vulnerable resource, essential to sustain life, development, and environment.... Water has an economic value in all its competing uses and should be recognized as an economic good.

—Dublin Principles of Integrated Water Resources Management, 1992

Chile's free-market Water Code turned 20 years old in October 2001. This anniversary was an important milestone for both Chilean and international debates about water policy because Chile has become the world's leading example of the free-market approach to water law and water resources management—the textbook case of treating water rights not merely as private property but as a fully marketable commodity. Other countries have recognized variations of private property rights to water, but none have done so in as unconditional and deregulated a manner as Chile. Because the 1981 Water Code is so paradigmatic an example of free-market reform, some people have praised it as an intellectual and political triumph, while others have criticized it as a social and ideological aberration.

The predominant view outside Chile is that the Chilean model of water management has been a success. Many economists and water experts in the World Bank, the Inter-American Development Bank, and related institutions share this view. Since the early 1990s, these proponents have used their considerable resources and influence to promote a simplified description of the Chilean model and its results, both elsewhere in Latin America and in the wider international water policy arena. Although they sometimes recognize flaws in the model, their general tendency has been to play down the importance of those flaws and instead to emphasize the model's advantages.

One of the major purposes of this book is to challenge that description and present a more balanced view of the Chilean model. More than 20 years of Chilean experience is a long enough period to show that the free-market model of water management has had marked weaknesses as well as strengths, and that the weaknesses are structurally connected to the strengths. In other words, both the strengths and the weaknesses are built into the current legal and institutional framework in Chile, in ways that have made them effectively impossible to separate. Thus Chilean water markets have worked fairly well in some important respects and under certain conditions, but in other contexts and for other purposes they have worked quite poorly. The problem areas include a range of critical water management issues, such as social equity, environmental protection, river basin management, coordination of multiple water uses, and resolution of water conflicts. These issues are at the core of current international debates and concerns about water policy and management.

Because Chile's Water Code did not seriously address these issues in 1981 (as explained in Chapter 2), it may be unfair to criticize the code for its failure to solve them later. But that is not the point here. The larger point is that after more than 20 years of experience, the current legal and institutional framework—which is determined by Chile's 1980 Constitution as well as by the Water Code—has shown itself incapable of handling these unforeseen problems. The current framework, as we will see, is characterized by a combination of elements that reinforce each other to maintain the status quo: strong and broadly defined private economic rights, tightly restricted government regulatory authority, and a powerful but erratic judiciary untrained in public policy issues and holding a narrow and formalistic conception of law. All indications are that these problems of water management will only worsen as the demands and competition for water increase, putting ever more pressure on the existing institutional framework.

A second major purpose of this book is to place the Chilean experience in a broader international and comparative context. Because of its symbolic importance to international water policy debates, the Chilean Water Code's recent anniversary presents a valuable opportunity to look back and assess the Chilean experience and to draw more general lessons about water policy reform. In this larger context, my goal is to argue for, and to demonstrate, a more qualitative and interdisciplinary approach to the *economics* of water. Such an approach emphasizes the institutional, legal, social, and political aspects of economic analysis—aspects that are too often missing or considered only superficially in conventional economic approaches.

The organization of the book reflects these two main purposes. Chapter 1 frames the international context of the Chilean case by reviewing recent international debates about the global "water crisis" and the growing need for major reforms of water law and policy. These reforms point toward what is called integrated water resources management. I focus particularly on the range of views about what it means to manage water as an "economic good," which has been a major theme in international water debates, and I ask which of the different economic perspectives on water management are truly compatible with an integrated approach. This is in some ways the fundamental question of the book, and it is a question that the Chilean experience is uniquely suited to help answer. Chapter 1 also gives some examples of the Chilean model's significance to other countries considering water policy reform.

The bulk of the book reviews the essential aspects of the Chilean experience. The discussion is based on years of research in Chile, including numerous interviews; analysis of legal, policy, and government documents; and academic publications, both Chilean and foreign.

Chapter 2 discusses the characteristics and background of Chile's 1981 Water Code, first by summarizing the law's major features and then backing up to describe its legislative and political history. This review focuses on the period since the 1960s, and especially the period of military dictatorship after 1973. The legal and political background is crucial to understanding the more recent issues examined in Chapters 3 and 4, which focus on the period since 1990. The broader theme of Chapter 2 is to underline the political nature of economic instruments—both the decision to adopt such instruments in general and the design of their specific features.

Chapter 3 looks at the evolving political and policy debate in Chile about how (or whether) to reform the Water Code, after the country's return to democratic government in 1990. Despite more than 13 years of sustained government efforts to pass some kind of legislative reform, strong political opposition has blocked most of the government's proposals.

Moreover, the terms of debate and the range of possible alternatives have narrowed dramatically. The broader message here is the overriding importance of the national political and institutional context for meaningful changes in water policy and management.

In Chapter 4 I review the progress in research and analysis about the empirical results of Chilean water markets. Like Chapter 3, this chapter focuses on the period from 1990 to the present, since before the transition to democracy, there was no research in this area. I discuss the areas of emerging consensus among both Chilean and foreign experts about how to describe the performance and results of Chilean water markets, and then I turn to the issues that have been neglected, including social equity, river basin management, environmental protection, and the resolution of conflicts among water users. More adequate study of these issues is essential because the available evidence shows that Chilean water markets have handled them poorly. Chapter 4 concludes by briefly describing recent and emerging water policy issues in Chile and pointing to directions for future research.

Finally, Chapter 5 presents overall conclusions about Chile's experience with its free-market water law, together with lessons for current international debates about water policy reforms. I summarize the strengths and weaknesses of the Chilean model of water management, both on its own terms and in comparative and international contexts. On the basis of that analysis I also return to my argument that the Chilean experience illustrates the need to encourage a broader and more interdisciplinary approach to water law and economics.

Although this book focuses on the specific issues of water resources, the general analysis is intended to apply to a much wider range of Latin American institutional and policy concerns. In particular, the Chilean experience of attempting to reform the 1981 Water Code is a superb example of what has come to be called second-generation or second-stage reforms. The first generation of reforms was the set of neoliberal economic policies implemented in nearly all Latin American countries from the 1970s through the 1990s—*neoliberal* being a synonym for "free market." These policies are often referred to as the Washington consensus because they have been aggressively promoted by the U.S. government and by the multilateral financial institutions based in Washington, particularly the World Bank and the International Monetary Fund. Neoliberal policies have been applied both at the level of national macroeconomics and within specific social and economic sectors (such as water resources in Chile).

Neoliberal policies have been highly controversial and have undoubtedly had both positive and negative impacts in different countries. Whatever one's perspective on these policies, since the mid 1990s even their strongest supporters have recognized the need to "reform the reforms"—that is, to address the problems of social inequity and institutional failure that have frequently resulted from too narrow an emphasis on privatization and free markets. Central to this second generation of reforms are efforts to reform judicial, legal, and regulatory systems.[1]

So far, however, very few examples of second-generation reforms anywhere in Latin America have enough of a track record to be evaluated empirically. In this respect, Chile's experience with water policy in the 1990s is unusual and has a resonance far beyond the specific arena of water issues.

CHAPTER 1

THE INTERNATIONAL CONTEXT: THE WATER CRISIS AND DEBATES ABOUT WATER POLICY

In the international context, the Chilean model of water management represents one response to what is increasingly recognized as a global "water crisis." This crisis has been caused by the confluence of several international trends in water use and water management—trends that are remarkably common and widespread throughout the world despite obvious differences in social, economic, and geographic conditions.

The core trend is the ever-growing demand for water for an increasing range of social, economic, and environmental purposes. These growing and multiplying demands are driven by long-term forces of population and economic growth. Continually increasing demands have made water resources relatively scarcer, which has raised water's economic value, intensified the levels of competition and conflict among different water users, and magnified the environmental impacts of water use. The dynamics of growing scarcity, economic value, conflict, and environmental impact reinforce one another and have led to a vicious cycle in many parts of the world.[1]

Moreover, water scarcity is not just a problem of water quantity; it also includes issues of pollution and water quality. Lack of sufficient water of adequate quality, for whatever purposes, is a problem of scarcity. The two aspects of water management are always physically interrelated.

Since the early 1990s the scope and severity of the world's water problems have received broad international recognition, as indicated by a series of high-profile international conferences. Especially prominent were the International Conference on Water and Environment, held in Dublin in 1992; the United Nations Conference on Environment and Development (also known as the Earth Summit), held in Río de Janeiro, also in 1992; and the Second World Water Forum, held in The Hague in 2000. These conferences and related events have led to international consensus about the need for major reforms of water laws, policies, and management to address the worsening problems of water scarcity and conflict. The Chilean model has often been mentioned at these international conferences as an example of successful reform.

Several themes have dominated international water policy debates. One is that water policy reforms should move in the direction of more "integrated" water resources management, including long-term environmental sustainability. This notion enjoys very broad rhetorical support.

A second theme is that water reforms should take a more "economic" approach, making use of market incentives and other economic instruments to increase the efficiency of water use and allocation. This approach is more controversial but is nonetheless fairly well accepted as a general principle, and it is the context in which the Chilean model is most commonly discussed.

A third theme is that water management should place more emphasis on problems of poverty and social inequity, especially in the poorer countries of the developing world.

Beyond the general rhetorical consensus, of course, there is much less agreement about what the specific policy reforms should be. At least in theory, integrated water resources management should include economic, social, and environmental issues. It is debatable, however, whether these kinds of reforms are really compatible.

In this book I focus most of my attention on the first two themes—the arguments for more integrated and more economic approaches to water management—and I fold the issues of social equity into that context.

INTEGRATED WATER RESOURCES MANAGEMENT

A great deal of ink has been spilled about integrated water resources management (often abbreviated IWRM), and I offer only a brief summary here. Integrated water resources management aims to be a comprehensive and interdisciplinary approach that recognizes and deals with the many social, economic, political, technical, and environmental aspects of water issues. According to the definition of the Global Water Partnership, "IWRM is a process which promotes the coordinated development and management of water, land, and related resources, in order to maximize the resultant economic and social welfare in an equitable manner without compromising the sustainability of vital ecosystems."[2] This requires understanding of the overall water cycle—that is, the continuous processes of water as it evaporates from the oceans, precipitates as rain and snow, and flows downhill across land surfaces and through rocks and soil back to the oceans.

Taking such a holistic perspective contrasts with the fragmented and sector-specific approaches that have historically dominated most countries' water laws, policies, and institutions. An integrated approach emphasizes the close connections between issues that are nearly always regulated separately, such as the relationship between water uses and land uses, the relationship between ground water and surface water, and the relationship between water quality and water quantity. Similarly, an integrated approach by its nature treats river basins and watersheds as the most appropriate geographic units for water management, rather than areas defined by political or administrative boundaries.[3]

One of the best-known expressions of the current international consensus about water management is the so-called Dublin Principles. These principles emerged from the International Conference on Water and Environment held in Dublin in 1992 (as a precursor to the Earth Summit in Río de Janeiro later that same year). The Dublin conference brought together many water experts and organizations from around the world and concluded with a public declaration of the need for more integrated water resources management, together with a description of four guiding principles to help achieve it. The Dublin Principles cover environmental, sociopolitical, gender, and economic issues, as follows:

> (1) Fresh water is a finite and vulnerable resource, essential to sustain life, development, and environment....
>
> (2) Water development and management should be based on a participatory approach, involving users, planners, and policy-makers at all levels....

(3) Women play a central part in the provision, management, and safeguarding of water....

(4) Water has an economic value in all its competing uses and should be recognized as an economic good.[4]

Although these principles are obviously very general, their concise expression has been useful in framing and publicizing the major issues in international water policy debates. Furthermore, this particular formulation has outlived the 1992 Dublin conference because of the subsequent creation of the Global Water Partnership (GWP), an international organization established in 1996 to promote and help implement the Dublin Principles around the world.[5]

It is evident that integrated water resources management is an ideal concept rather than a set of specific guidelines and practices. Like "sustainable development," it is a phrase that can mean all things to all people; because it seems to mean everything, it may end up meaning nothing. Nevertheless, even as a rhetorical term IWRM is significant as a reflection of the current international consensus about general principles of water management, particularly the need to take into account a wide range of factors, professional disciplines, and analytical perspectives.

There are also substantive parallels between IWRM and sustainable development, since achieving either goal depends on meeting the same three essential criteria: economic efficiency and growth, social equity, and environmental protection. These three criteria are often referred to (and graphically portrayed) as the triangle of sustainable development.

The holistic and interdisciplinary nature of integrated water resources management is also illustrated by the recent surge of interest in "water governance." Because it encompasses the broad social and political aspects of water management, water governance goes beyond the scientific and technical aspects that have traditionally been the emphasis of water management. The Global Water Partnership has declared that "the water crisis is mainly a crisis of governance"[6] and has defined water governance as "the range of political, social, economic, and administrative systems that are in place to develop and manage water resources, and the delivery of water services, at different levels of society." This definition draws on a more general definition of governance by the United Nations Development Program: "Governance is the exercise of economic, political and administrative authority to manage a country's affairs at all levels ... it comprises the mechanisms, processes, and institutions through which citizens and groups articulate their interests, exercise their legal rights, meet their obligations, and mediate their differences."[7]

Governance is a very broad and inclusive concept whose aspects are inherently difficult or impossible to quantify, and it has generated a growing literature in various fields of policy and regulation. In this book I do not analyze this literature in more detail. Nevertheless, it is clear that issues of governance are closely related, if not identical, to institutional arrangements, as discussed in the following sections.

PERSPECTIVES ON WATER AS AN "ECONOMIC GOOD"

The most controversial of the Dublin Principles has been the fourth one—that water "should be recognized as an economic good." What does this mean, and what are the policy implications? Most of the international debate about these questions has focused on the advantages and disadvantages of free markets and privatization.

Many people agree that the economic efficiency of water use and allocation is very important and that market forces and economic incentives, including prices, are powerful tools for increasing that efficiency. They disagree, however, about how free or unregulated such markets should be. The strongest market advocates argue that managing water as an economic good means treating water as a fully private and tradable commodity, subject to the rules and forces of the free market; from this perspective, the economic value of water is the same as its market price. The extreme opposing viewpoint considers access to water a basic human right and sees market forces and prices as unacceptable or irrelevant. In fact, the Dublin Principles try to have it both ways. Immediately following the controversial sentence—"Water has an economic value in all its competing uses and should be recognized as an economic good"—the statement continues: "Within this principle, it is vital to recognize first the basic right of all human beings to have access to clean water and sanitation at an affordable price."[8]

An intermediate position is that water should be recognized as a scarce resource, which means that the available supplies are insufficient to satisfy all demands and that trade-offs are therefore necessary in allocating water to different uses. These trade-offs, however, need not be made via private or unregulated markets.

It is important to make a basic distinction here between the privatization of water *resources* and the privatization of water *services*. The privatization of water resources involves the ownership, use, management, and regulation of water resources themselves—the focus of this book. In contrast, the privatization of water services involves the organizations and infrastructure that supply water to consumers, and the costs and conditions of how that water is delivered. Although the two areas are closely related and are often lumped together when people talk loosely about the "privatization of water," the core issues and objects of concern can be quite different. Much of the recent international controversy in this area—for example, at the Second World Water Forum in The Hague in 2000—has been about the provision of water services rather than about natural resources or environmental issues per se.[9]

In this chapter I review the specific interpretations of the fourth Dublin Principle that have been offered by international water economists. Before I do that, however, I want to raise briefly some general issues about the varieties of economic perspectives and schools of thought.

For the purposes of this book, I distinguish between what I will call "narrow" and "broad" perspectives on economic analysis. The distinction is based on two essential characteristics: first, how open the analysis is to academic disciplines and methods other than orthodox neoclassical economics; and second, how much attention is given to the institutional framework of markets and of economic activity in general. These two characteristics are closely related—indeed, they are in a sense two ways of expressing the same idea, since institutional arrangements and sociopolitical context are typically outside the realm of neoclassical economics, by definition. (When I use the terms *institutions* and *institutional arrangements*, I refer broadly to the legal, political, and social norms, rules, and organizational structures that shape the patterns of human behavior.)

In general terms, when I refer to narrow economic perspectives, I mean the more formal, quantitative, and technical approaches to neoclassical economics. The narrow perspective also includes neoliberal economics, which is the more extreme free-market version of neoclassical economics. I recognize that many neoclassical economists are not neoliberals, but even so, their views of institutions and of the boundaries of economics are very similar. In contrast, when I refer to broader economic perspectives, I mean approaches that rely more on qualitative, historical, and interdisciplinary analyses, such as the fields of institutional economics, political economy, and ecological economics.[10] Later in this chapter I will mention some examples of these approaches as applied to water problems.

Another way to express the contrast between broad and narrow perspectives is to compare how much the realm of "economics" is considered separable from history and other social sciences. From a narrow perspective, economic methods are essentially mathematical, and economic analysis is largely independent of social, political, historical, cultural, or geographic factors. These factors are notoriously hard to quantify, and hence they are generally put aside or assumed as given.

I do not wish to exaggerate the distinction between narrow and broad economic perspectives, and I do not want to criticize neoclassical economics unfairly. The different perspectives are located along a continuous spectrum, rather than being polar opposites, and there are many neoclassical economists who are somewhere in the middle. (This is especially true of older generations of economists.) I myself rely on some of the core principles of neoclassical analysis (although not the math), as will be clear throughout much of this book, and I fully recognize the benefits of markets. Furthermore, even economists at the narrower end of the spectrum generally recognize, at least in theory, the importance of legal, institutional, and political arrangements to how markets work. The problem is that these arrangements are still considered "noneconomic" conditions, whose prior existence must be assumed in order to carry out quantitative economic analysis.

An institutional economic perspective, in contrast, focuses directly on the institutional arrangements, even though this requires qualitative analysis of noneconomic factors, such as law, politics, culture, and historical and social context. Since the rules governing markets come before the markets themselves, using market-based methods to analyze the rules can lead to conclusions that are too narrow, if not simply mistaken. Instead we should analyze these rules in terms of the values and interests of the people who influence their design.

Although institutional economics is to some degree a critique of neoclassical economics, the larger objective is to build on and incorporate the insights of neoclassical analysis rather than to reject them. My point here is simply to emphasize the limitations of neoclassical theory when applied to institutional matters. In recent decades there has been a great deal of academic work in the fields of "law and economics" and the "new institutional economics," in which neoclassical economic analysis has been applied to a wide range of legal and institutional issues. Property rights have been a particular center of attention, since property is one of the fundamental areas of overlap between economics and law. Although economists' increased attention to these issues has been an important change, the depth of the insights achieved has often been rather limited from the point of view of someone trained in law, politics, history, or sociology. It is still frequently the case that economic analyses oversimplify or misunderstand the nature of law and institutions—that is, what they consist of, how they work, and how they are shaped by social and political context. A common example of this problem is the assumption that the main economic purpose of legal institutions is to facilitate markets. In light of these oversimplifications, many economists' declarations about the importance of effective institutions can seem to be ritual incantations, without much substantive impact on their analysis.[11]

All of these issues have political as well as academic and analytical implications. The term *institutions* is, in part, a synonym for "government" or "the state," and therefore how people approach the institutional aspects of markets is directly related to their views about the proper role of government regulation.

It is in this context of different economic perspectives that I am frequently critical of the World Bank's positions on water and related issues. The Bank itself is not my principal target or concern in this book, except to the extent that it represents a narrow economic perspective. That perspective has unfortunately been so dominant that World Bank analyses of water laws and institutions have tended to be rather superficial, and at worst ideological or dogmatic. This is a problem even with the Bank's better publications, such as the 1993 *Water Resources Management*.[12]

That policy paper, published the year after the Dublin Conference, is a relatively balanced discussion of integrated water management. On the one hand, it calls for more use of market mechanisms, privatization, decentralization, and efficient pricing, along with greater participation by users. On the other hand, it recognizes the need to adopt a "comprehensive analytical framework" based on an intersectoral approach and including strong legal and regulatory institutions. Unlike the Dublin Principles, however, the Bank's priority is unmistakably markets and market-oriented policies, which are examined and recommended in some detail. In contrast, the discussion of the comprehensive analytical framework and regulatory institutions is more abstract and rhetorical and offers little guidance to anyone trying to apply the policy recommendations.

Those limitations are also evident in other publications by prominent international water economists working at or closely associated with the World Bank. Such work (to which I will return later) illustrates the disappointing results of applying conventional economic wisdom to the legal and institutional aspects of water resources management. For example,

Mark Rosegrant and Hans Binswanger have argued strongly in favor of markets in tradable water rights in developing countries. Among their many claims about the benefits of such markets, one of the most relevant to integrated water resources management is their assertion that "appropriately defined property rights will in many instances cause farmers to internalize and thereby eliminate externalities." However, they say nothing more specific about what the "appropriate" definitions of property rights might be—other than, of course, tradable. The authors mention the importance of the "real-world institutional and technological context of developing-country irrigation," and they also mention the need to reform laws and institutions, but again without offering any further explanation or details about what might be involved.[13] Hence their assertion, quoted above, may be true in economic theory, but as a guideline for legal and institutional practice, it offers very little. Moreover, the article focuses on irrigated agriculture, so even if we put aside the analytical weaknesses, the argument has only limited application to more complex and intersectoral issues of water management.

A second example is a recent book presenting an overview of the theory and practice of water markets around the world, a collection edited by K. William Easter, Mark Rosegrant, and Ariel Dinar. In the book's introductory and concluding chapters, the editors organize their analysis around the basic conceptual distinction between what they call formal and informal water markets. The main difference is how water rights trades are enforced: either within the "legal or administrative system," in the case of formal markets, or by water users themselves without access to those systems, in the case of informal markets. Formal markets, in other words, are regulated by government. Because "the scope of informal markets is likely to be limited," the authors argue that the most significant and complex issues in water management—such as trades between economic sectors, interbasin water transfers, and externalities—must instead be handled by formal markets.[14]

These authors are not free-market radicals. They emphasize that "for formal water markets to work, the government will have to take an active role in establishing the appropriate institutional and organizational arrangements." They present a list of "strategies for mitigating problems and constraints" of water markets. Many of these strategies have major political, distributional, and administrative implications, which in many countries would be highly controversial or very difficult to implement in practice, or both.[15] (In Chile, for example, as we will see, most of the proposed mitigating strategies would be either unconstitutional or politically infeasible and even in the best of political circumstances would require significant improvements over existing institutional capacity.) Easter et al., however, do not discuss these political and institutional difficulties in detail. Instead, they note the kinds of government regulation that would be required in theory and then proceed to their conclusion: "Contrary to the 'nay-sayers,' water markets have worked and can be a superior mechanism for reallocating water."[16]

The problem here is not that the arguments of these experts are unreasonable. The problem is that their arguments rest on assumptions that are rarely met in practice, and this makes it dangerous to treat the arguments as policy recommendations. Easter, Rosegrant, and Dinar are all senior, experienced water economists. Nonetheless, whatever professional or personal appreciation they may have about the complexities of government or the policy process, their written arguments suggest that invoking a set of assumptions is, by itself, sufficient to get water markets to operate effectively.

In short, the differences between broader and narrower economic perspectives are important because of their implications for legal and institutional arrangements. The implications are critical to answering the question raised earlier: How does recognizing water as an economic good relate to the larger challenges of integrated water resources management? However generally IWRM may be defined in theory, if it is to mean anything in practice, it must include the following essential functions:

- coordinating different water uses, upstream and downstream, at the level of river basins and watersheds;
- resolving water conflicts;
- internalizing or otherwise dealing with economic and environmental externalities;
- defining and enforcing property rights; and
- monitoring compliance with the rules about water use and management.

All of those management functions are closely connected, and carrying them out effectively depends on the same general legal and institutional framework. It also necessarily involves distributing the costs and benefits of different water uses among different people and groups in society, and hence these processes cannot be politically neutral or purely technical.

One final comment is needed about my own disciplinary approach. By placing so much emphasis on the importance of law and institutions in matters of markets and economics, I do not focus only on their formal aspects—the "law on the books." Instead, my primary concern is with law in its social context—the "law in action." Hence I look at legal rules, processes, and institutions through the lenses of history and the social sciences, in keeping with the "law and society" academic tradition in the United States, which considers the distinction between laws on the books and laws in action a basic principle. In short, I take the same interdisciplinary approach to law that I argue should be taken to economics.[17]

When we combine these two approaches, we can move toward both a richer understanding of the real world and a more grounded approach to public policy. I refer to this combination as comparative law and economics because a comparative approach is inherently interdisciplinary, qualitative, and historical. This approach is comparative even if applied to a single country, since it examines that country in relation to the broader international context.

ECONOMIC INTERPRETATIONS OF THE FOURTH DUBLIN PRINCIPLE

In recent years prominent water economists have tried to explain the meaning of recognizing water as an "economic good" to diverse water policy audiences, including economists as well as noneconomists. In this section I review examples of these explanations, selected for several reasons: they represent different points along the spectrum from narrow to broad economic perspectives, these publications have been frequently cited in recent international water policy debates, and their authors are well-known in international water policy circles or work at international organizations that are influential in water management issues.

I pay special attention to the authors' uses of particular terms and definitions, which vary a great deal. This is not merely a semantic point. Besides the obvious difficulty that these differences pose to readers trying to understand the basic concepts and arguments—particularly readers from different disciplinary backgrounds—the differences indicate the fundamental assumptions underlying the authors' theoretical frameworks. For example, the meaning of a term like *economic value* is far from being as simple or self-evident as it may seem. These basic assumptions and definitions, whether explicit or implicit, are often where the action is, especially when it comes to policy implications.

A useful starting point is the work of John Briscoe, a senior water adviser at the World Bank who has played a visible role in articulating and presenting the Bank's viewpoint and policies at national and international meetings. Given the Bank's great influence in international water management, both in setting the terms of debate and in determining the criteria for public and private investment, Briscoe's role has been of strategic importance.[18]

Briscoe has presented a fairly simple analytical framework for managing water as an "economic resource," as he calls it. He argues clearly and forcefully in favor of markets and neoclassical economics. (He also strongly favors the Chilean model, as discussed later.) He begins by distinguishing three components of water as an economic resource: value, use cost, and opportunity cost. He then analyzes the interaction among these three components in different water sectors, focusing particularly on the contrast between urban water supply and agricultural irrigation.[19]

The core of Briscoe's argument is that there is, or should be, an equation between the value of water on one side and the costs of water on the other. He defines *value* to mean simply "willingness to pay," following the basic conventions of neoclassical theory. He does not mention any alternative concepts or definitions, although he does discuss the methodological difficulties of estimating water values and prices. The economic cost of water, on the other side of the equation, is composed of both use costs and opportunity costs. The meaning of *use cost* is straightforward: it is the cost of "constructing and operating the infrastructure necessary for storing, treating, and distributing the water." Opportunity costs are much harder to determine, since they are the costs "incurred when one user uses water and, therefore, affects the use of the resource by another user." Because opportunity costs refer to the alternative uses for water resources, they are also closely related to conflicts among water users.[20]

Briscoe illustrates his argument and draws policy implications by comparing the relative values and costs of urban water supply (in both developed and developing countries) with those of irrigated agriculture (in both privately and publicly financed systems). The contrast is striking. Urban water supply is a low-volume, high-value use whose use costs are relatively high (per unit of water) but whose opportunity costs are low because alternative water uses (particularly for agriculture) are typically less valuable. Irrigation, on the other hand, is a high-volume, low-value use whose use costs are lower (per unit of water) but whose opportunity costs are higher because alternative, nonagricultural water uses are more valuable. Therefore, Briscoe argues, in the agricultural sector, "from the point of view of management of water as an economic resource, the key challenge is to ensure that users take the opportunity cost of water into account." In the urban sector the situation is reversed: "Ignoring opportunity costs is thus a matter of minor practical importance when it comes to the economic management of urban water supplies, but a matter of huge practical significance when it comes to irrigation."[21]

Although Briscoe's framework is clear and useful in explaining some important concepts, it has serious limitations when applied to integrated water resources management or to problems of river basins. The analysis is almost entirely focused on urban water supply and irrigation, both of which are consumptive uses. There is no discussion of environmental water uses, protection of water quality (except as part of the costs of urban water supply), nonconsumptive water uses (such as fishing, hydropower, or navigation), or how to coordinate all these different water uses. Externalities are addressed in a very limited manner, although in theory they are reflected in opportunity costs.

In a subsequent paper, Briscoe extends his framework to take account of externalities, both positive (e.g., return flows) and negative (e.g., pollution). "These externalities," he asserts, "are easily incorporated into the conceptual framework ... by simply increasing the supply costs to include the cost of mitigating the negative externalities."[22] This statement is a good illustration of the dangers of a narrow economic perspective, for though it may be conceptually true, it does not recognize the many political and institutional difficulties that must be overcome when people try to internalize externalities. Even Briscoe's extended framework, in other words, is too narrow to include environmental or social factors—as he himself tacitly acknowledges by listing these factors as separate from the idea of water as an economic good.[23]

A similar framework is presented by Peter Rogers, Ramesh Bhatia, and Annette Huber in their effort to explain to noneconomists the concept of water as an economic good. Their paper was one of the "background papers" written for the Global Water Partnership to promote and make operational the Dublin Principles of integrated water resources management (as described earlier this chapter). Like Briscoe's work, their basic explanation is fairly clear and follows a technical approach based on neoclassical economics and engineering. In light of the GWP's broader mission, the paper's lack of attention to social, political, or institutional matters is disappointing.[24]

Rogers et al. present an analytical framework that, like Briscoe's, puts economic costs on one side of the equation and economic value on the other, and they argue that sustainable use of water requires that the two sides be balanced. Their framework has a more elaborate breakdown than Briscoe's of the different components of cost and value. On the cost side of the equation, they define the "full cost" to comprise (1) the "full supply cost" (i.e., capital charges plus operation and maintenance, or what Briscoe calls use cost); (2) opportunity costs; and (3) economic and environmental externalities. Their "full economic cost" includes all of the above except environmental externalities. They recognize that it is often hard to separate economic externalities from environmental externalities, but they insist on the distinction because "environmental externalities are usually inherently more difficult to assess economically."[25]

It is on the value side of the equation that Rogers et al. diverge more from Briscoe. They define water's "value in use" as the sum of two components: economic value and intrinsic value. Economic value in turn has four components: (1) "value to users of water" (defined as marginal value of product for industrial and agricultural uses, and as willingness to pay for domestic use, or what Briscoe calls simply value); (2) "net benefits from return flows"; (3) "net benefits from indirect use" (i.e., various positive externalities); and (4) "adjustment for societal objectives" (i.e., nonmarket criteria such as distributional equity).[26] Their inclusion of nonmarket social criteria in economic value reflects a broader perspective than Briscoe's.

The discussion of "intrinsic value," however, is brief and incomplete. Rogers et al. recognize the validity of noneconomic concepts of value, and they refer loosely to the "benefits occasioned by environmental management,"[27] but they essentially abandon further analysis in the face of the difficulties of measuring or estimating these values. Despite their limited description, intrinsic value seems to be the only part of their framework that might include environmental or ecological values. (The authors also mention "reliability of water supply" and "water quality" as two issues that affect cost and value but nevertheless do not alter the basic framework.)

The framework proposed by Rogers et al. is limited in several ways. First, their analysis of externalities can be confusing. In their model, both economic and environmental externalities are located on the cost side of the equation; these are presumably negative externalities, although they are not identified as such. At the same time, several components on the value side also include externalities, though they do not seem to account for this. According to their definition of economic value, for instance, both kinds of "net benefits" (i.e., benefits from return flows and benefits from indirect use) essentially refer to positive externalities, although they do not use the term. In addition, in many circumstances the "adjustment for societal objectives" is at least partly an effort to compensate or redistribute the impacts of externalities. Moreover, to the extent that "intrinsic value" includes all environmental (noneconomic) costs and benefits, presumably it must include environmental externalities in some way. But Rogers et al. say nothing about how, or whether, these externalities of cost and of value are related to each other across the two sides of the equation.

A second major problem is that environmental factors and values are poorly incorporated into this framework, as the authors themselves admit.[28] They suggest no remedy other than econometric techniques of nonmarket valuation (which, whatever their academic merits, have debatable applicability to public policy). This reflects the paper's overall disciplinary perspective: the framework is strong on technical and quantitative fields like neoclassical economics, engineering, and systems analysis, but weak or silent on law, politics, or other social sciences. Institutional aspects of water management, in consequence, are barely mentioned.

Other development economists have argued for a broader perspective on the meaning of water as an economic good. A good example is a paper written by C.J. Perry, David Seckler, and Michael Rock, and published by the International Irrigation Management Institute. Although Perry et al. focus particularly on water use in irrigated agriculture, their analysis is more generally applicable. They interpret the phrasing of the fourth Dublin Principle as having been deliberately vague in an effort to find a compromise between the two extremes of free-market versus antimarket viewpoints—a compromise, in other words, "between those, mainly economists, who wanted to treat water in the same way as other private goods, subject to allocation through competitive market pricing, and those who wanted to treat water as a basic human need that should be largely exempted from competitive market pricing and allocation."[29] Perry et al. consider both extremes dogmatic and simplistic and therefore dangerous when applied to public policy. They criticize Briscoe's analysis, discussed above, as an example of promarket bias that oversimplifies the larger issues of social welfare.

Perry et al. argue that water is undoubtedly an economic good simply because it is scarce relative to its many alternative uses. Their crucial point, however, is that even in orthodox neoclassical terms, water can be both a private and a public good, depending on the context. In contexts where its features as a private good are dominant, water can be allocated through market mechanisms with good results. In the many circumstances where its features as a public good are more important, water is subject to the classic problems of "market failure," such as externalities, transaction costs, monopoly power, and insecure or ineffective property rights, and in such circumstances nonmarket institutions are essential.

The distinction between private and public goods is so elemental in neoclassical economics, and the recognition that water shares features of both is so common, that it is surprising that this point is often obscured in international water policy debates (including the papers by Briscoe and Rogers et al. discussed above). Along the same lines, when they discuss the value of water, Perry et al. draw a distinction between "financial value," which can be readily measured by market prices, and the broader category of "economic value," for which market prices alone are a poor measure.

In conclusion, Perry et al. describe a set of legal, technical, and institutional preconditions that should be met before market forces should be introduced into water allocation. These preconditions include the definition of all users' entitlements "under all levels of resource availability"; the existence of infrastructure to deliver these entitlements and acceptable standards for measuring delivery; the availability of "effective recourse" to users who do not receive their entitlements as well as to third parties who are affected by changes in use; the payment of user fees; and regulatory agencies capable of reviewing and modifying large-scale water transfers. The authors are well aware that meeting all of these preconditions is not easy. They conclude that "absent these basic prerequisites—the norm in most developing countries—the more extreme variants of privatization, such as full water pricing and unregulated market allocations, are likely to do more harm than good."[30]

Desmond McNeill, another development economist and former water adviser to the Norwegian Agency for Development Cooperation, offers a similar argument. He laments the "confused and heated" nature of the international debate in this area, and like Perry et al. he states that water is obviously an economic good simply because it is a scarce resource for which there are many competing uses. Taking this position, however, "does not necessarily imply that a 'market price' must be paid for [water], or even that it must be paid for at all. It means simply that water is ... a valuable resource that should not be wasted."[31]

McNeill argues that the assumption that treating water as an economic good requires market pricing reflects a narrow "economistic" perspective rather than a broad "economic" perspective. In his view, the economistic perspective is strictly neoclassical and confined to the "domain of the market," in which the overriding objectives are economic efficiency and growth. The economic perspective, on the other hand, is broad enough to bring together economistic objectives with social and ecological objectives, as required by the triangle of sustainable development. Although McNeill criticizes the economistic perspective for failing to take those other objectives into account, he is equally critical of the common argument that water is a "social good"—a basic human need—rather than an economic good. He agrees that the argument reflects valid concerns about public health and equitable access to drinking water but says it leads to confused analysis and bad policymaking if people refuse to acknowledge that resources are scarce and that trade-offs must be made.[32]

McNeill points out that this whole debate must be understood in the context of the politics of international organizations. There are different sectoral interests at stake as well as different disciplinary perspectives. People and organizations whose primary concerns are drinking water supply or agricultural development, especially in poorer countries, are likely to resist arguments that lead to higher water prices.

A last example of a senior water economist with a broad perspective is F. Lee Brown, of the University of New Mexico. Like Perry et al. and McNeill, he argues that recognizing water as an economic good simply means that "it is scarce relative to the uses to which it can be put." This recognition, Brown says, "is not novel to the 1990s despite the emphasis on the term provided by the Dublin statement. Water has been recognized and treated as an economic good by every civilization that encountered a scarcity of it .... What is new and different about the modern perspective on water is not that it has *economic* characteristics but that it is acquiring the status of a commercial *commodity*, distinct and divorced from its value to the community in many ways that differ from its economic scarcity value alone."[33] Brown argues that the way to merge the commodity perspective on water's value with the community perspective is to focus on the concept of trade-offs, rather than on prices or markets.

STRETCHING THE NEOCLASSICAL PARADIGM: INSTITUTIONAL AND ECOLOGICAL ECONOMICS

At the broader end of the spectrum of approaches to economic analysis are two schools of thought that challenge the dominant neoclassical framework and are generally overlooked or dismissed by neoclassical economists: institutional economics and ecological economics. I want to close this chapter's theoretical overview by summarizing two examples of work on water management that illustrate these perspectives. As mentioned earlier, I am referring here to the "old" rather than the "new" institutional economics (the latter being the application of neoclassical economic analysis to such legal and institutional issues as property rights).

The first example is a classic paper, written in 1982, by the institutional economist Daniel Bromley, in which he discusses the limitations of certain fundamental principles of neoclassical theory when it comes to recommending public policies for natural resources management. Bromley undermines the common assertion that neoclassical economic analysis is "scientific" or value-neutral; he advocates an institutional perspective instead.[34]

Bromley argues that the concept of economic efficiency—in many ways the core concept of the neoclassical approach—has major shortcomings as a guiding principle for policy decisions. In the first place, the relative efficiency of different allocations of resources can be determined only for a given set of institutional arrangements and a given initial distribution of resources. The logic cannot be simply reversed: efficiency cannot be the standard for comparing institutional arrangements. In the second place, choosing among institutional arrangements almost inevitably affects distributional equity—that is, the distribution of the costs and benefits of resource uses among individuals and groups within society. From Bromley's perspective, since economists must assume certain social and institutional conditions in order to calculate economic efficiency, the resulting calculations have some built-in bias. Economic efficiency is undoubtedly an important principle to consider in making public policy decisions—decisions that typically involve choosing among institutional arrangements—but it should not be understood as essentially technical or politically neutral.[35]

Bromley emphasizes the essential role of law and legal institutions—courts, legislatures, and administrative agencies—in determining economic values. These institutions make and enforce the rules that determine people's rights, duties, and relationships. Markets and prices are an important part of the process of determining and measuring value, but in crucial respects they are the *effect* of institutional arrangements, not autonomous mechanisms. Bromley concludes by posing "three fundamental economic questions" concerning natural resource use: Who makes the rules? Who receives the benefits? Who pays the costs? These questions underline the inherently institutional and distributional character of natural resource economics.[36]

A second example is the work of Federico Aguilera, a leading Spanish water economist who works in both institutional and ecological economics. In a 1998 paper discussing the elements of a "new water economics," Aguilera describes a contemporary historical process in Spain that is very similar to the experience of recent decades in the United States. He describes the end of the previous era of Spanish water policies, which focused entirely on building and subsidizing water infrastructure to increase the available supplies, and the beginning of a new era that emphasizes water management instead of development.[37]

Aguilera compares three economic perspectives on water as a resource. Two are traditional neoclassical concepts: first, the notion of water as simply a factor of production, with little attention to the institutional framework or rules that determine how water is used; and second, the notion of water as a financial asset, which can be managed or disposed of according to its profitability and risk in relation to other financial assets. The third perspective is the notion of water as an "ecosocial asset," which means "the capacity to satisfy a whole collection of economic, social, and environmental functions, both quantitative and qualitative."[38] This third perspective includes the notion of water as a factor of production but is incompatible with the notion of water as a financial asset.

Aguilera argues that Spain is currently in a difficult and conflictive transition toward the third economic perspective—the notion of water as an ecosocial asset. The first step in this transition is to emphasize management of water demand rather than seek to develop new supplies. This transitional phase includes growing social awareness of environmental problems, growing social conflicts and public participation, and growing use of incentives for more efficient water use and distribution. The following step will be to achieve "integrated management of water and territory," which will require both a new economics and a new culture of water.[39]

In short, Aguilera's broad perspective on water economics brings us back to the international debates about water crisis and water policy reform. As discussed earlier in this chapter, moving toward more integrated water resources management will require a comprehensive and interdisciplinary approach that strengthens the emphasis on long-term environmental sustainability.[40]

INTERNATIONAL SIGNIFICANCE OF THE CHILEAN MODEL

The international significance of Chile's 1981 Water Code is due to its unique and extreme free-market approach to the issue of managing water as an economic good—an approach that will be described in detail in the next chapter. Since the 1990s, the fame of the Chilean model—usually referred to simply as "Chilean water markets"—has spread among international and Latin American water experts. Much of this fame is due to the World Bank, which has actively publicized the Chilean case as a model of success and an inspiration for water policy reforms in other countries.[41]

Evidence that the Chilean Water Code has been considered an international model comes in different forms. One is the steady stream of publications by the World Bank itself or by economists associated with the Bank. (I will discuss the substantive arguments of these publications in Chapter 4; here I want simply to note their role in publicizing the Chilean case.) For example, the Chilean case featured prominently in a 1994 article by Rosegrant and Binswanger—economists at the International Food Policy Research Institute and the World Bank, respectively—who present a comprehensive argument in favor of establishing markets in tradable water rights in developing countries (as discussed earlier in this chapter). That same year the World Bank published a report, written by Bank economist Mateen Thobani, that recommended promarket water law reforms in Peru and relied heavily on a description (albeit selective) of the Chilean law. In 1995 the Bank published two technical papers about Chilean water markets, which have been widely cited. Subsequently the Chilean case has been routinely included in international overviews, such as the World Bank's *Water Markets in the Americas* and a 1998 edited book about water markets around the world (also cited earlier), two of whose chapters are dedicated to Chile.[42]

United Nations agencies concerned with comparative water law reforms have also highlighted the Chilean case, although generally from a more critical perspective than the model's proponents. Since the mid 1990s, as I discuss in Chapter 4, the UN Economic Commission for Latin America and the Caribbean has been especially active in this respect, publishing several critical analyses of the Chilean Water Code. These analyses have had the explicit goal of counteracting the glowing reports of the Chilean model that have been circulated throughout Latin America.[43]

A second piece of evidence is the variety of other countries that have studied and considered following the Chilean example since the early 1990s, as part of their own national processes of discussing water policy reforms. Within Latin America alone these countries have included the following:

- Mexico (which passed its own new water law in 1992, incorporating market instruments but in a more balanced and pragmatic manner than the Chilean model);
- Peru (which has not yet adopted a new water law, after nearly 10 years of on-and-off political debate in which the Chilean model has been prominent);[44]
- Bolivia (where the recent "water wars" about privatizing water services in Cochabamba also involved new water rights legislation that was partly inspired by the Chilean model);
- Argentina (particularly the Province of Mendoza, just across the Andes from Chile, which has Argentina's largest area of irrigated agriculture and the longest and most sophisticated tradition of water law and water resources management); and
- Nicaragua and El Salvador (both of which have recently considered passing their first modern water rights legislation).

Many of those countries felt explicit pressure to follow the Chilean model from the World Bank or the Inter-American Development Bank. Outside Latin America, countries that have examined the Chilean model as part of their own recent processes of water policy reform include Australia, Indonesia, South Africa, Spain, Taiwan, and Vietnam.[45]

So far, no other country has actually copied the Chilean Water Code, although several of the countries named above have come close. By the late 1990s the Chilean model had become more controversial in international water policy circles, as some of its problems became better known. Many people have wondered whether such a model could ever be approved by a democratic government (as it was not in Chile). Nonetheless, the Chilean model continues to be a common point of reference in international water policy debates, and its proponents continue to praise its advantages even as they have been forced to recognize its weaknesses.

The fame of Chilean water markets among international water experts raises crucial questions.

- First, how successful have those markets really been, in terms of concrete results? This is the question that has attracted by far the most international attention, as I describe in Chapter 4.

- Second, what is the relationship between water markets and other critical water management issues in Chile? In other words, whatever the results of Chilean water markets, what have been the consequences of such promarket legal and institutional arrangements for dealing with water problems other than allocation of resources?

The second question has attracted much less attention, although in my view it is ultimately more important. In this context the Chilean experience offers a unique opportunity to examine the fundamental issue about the Dublin Principles that I raised earlier: *Is the free-market approach to recognizing water as an economic good, like the narrow economic perspective associated with that approach, compatible with the broader and long-term goals of integrated water resources management?*

According to proponents of the Chilean model, the answer to this last question is clearly yes. The World Bank's John Briscoe, for example, whose analysis of the fourth Dublin Principle was described above, has publicized Chilean water markets as a prominent example of "good practice" in managing water as an economic good.[46] He is so enthusiastic about the Chilean case that he abandons his generally sober and technical tone to praise the water market as "a *brilliant conceptual solution* to the enduring problem of reconciling practical and economic management of water .... This is the *genius of the water market approach*—it ensures that the user will in fact face the appropriate economic incentives" (i.e., opportunity costs rather than the costs of water storage and delivery, which are the usual basis of water prices). He argues that "in well-regulated river basins in arid areas of Chile, the water markets function as one would wish," though he fails to mention that in fact there is only one such basin in the entire country.[47]

Briscoe recognizes that basin-level management institutions are absent or ineffective in Chile and that river basin issues and water conflicts present major challenges. In fact, he describes the Chilean case as "good practice" for dealing with problems of water *scarcity* rather than water *quality*. For water quality management he prefers the French model, which is very different and involves a level of government intervention that would not be politically feasible in Chile.[48] In some of his publications, Briscoe qualifies his praise of the Chilean model with more balanced comments about its limitations and problems and about the current Chilean government's attempts to address them. Nevertheless, his overall enthusiasm for the model leads him to conclude that the "system of tradable water rights and associated water markets *is a great achievement and is universally agreed to be the bedrock on which to refine Chilean water management practices.*"[49] As we will see in Chapter 3, that same "bedrock" has made it extremely difficult to implement later "refinements," and the agreement within Chile is far from universal.

Upbeat assessments continue to be influential in international water policy circles. In a recent paper about water governance for the Global Water Partnership, Peter Rogers and Alan Hall describe the Chilean case as "*a world leader in water governance.*" They recognize that "many mistakes with openness, transparency, participation, and ecosystem concerns were made in the hurry to get effective water markets established," but nonetheless they conclude that "the system is adaptive and now these concerns are being addressed 20 years after the initial laws were passed."[50] Like Briscoe, however, their optimism about the prospects for correcting the Water Code's flaws is based on their ignorance of Chile's political and constitutional system. As I will show, the system is in fact not adaptive at all.

In this book I argue that the positive assessments of the Chilean Water Code are simplistic or mistaken in several crucial respects, and in consequence it is misleading and dangerous to present the Chilean case as a model of success whose few problems other countries can avoid. Although the Chilean model undoubtedly has some important advantages, these advantages are structurally tied to disadvantages that are at least as important and that should receive equal attention. Beyond the specific details of the Water Code's pros and cons, the Chilean experience offers broader lessons by demonstrating the critical limitations of a narrow approach to water economics and the failure of such an approach to adequately address the legal and institutional arrangements that are essential to integrated water resources management.

WATER MARKETS AND WATER POLICY IN OTHER COUNTRIES

There is an abundant literature about water markets in many parts of the world, apart from the specific debate about what it means to "recognize water as an economic good." In this book I do not provide an extensive review of that literature, but a few general comments may be useful here.

The term *water market* is often used loosely. Despite some general similarities, the specific issues and objectives involved can vary a great deal by country and region. For example, water markets in India and Pakistan are primarily for groundwater use, involve peasant farmers, and function apart from, and often complementary to, canal systems for surface water (which are owned and managed by government agencies).[51] Water markets in Australia, in contrast, often involve interstate rivers and problems of water quality, including salinity and maintenance of in-stream environmental flows.[52] In Spain, proposals to implement water markets have been driven by political debate about reducing subsidies for irrigation and avoiding the construction of new dams and reservoirs, in the context of European Union agricultural and environmental policies. In the Canary Islands, however, which are part of Spain, existing local markets for groundwater are strongly affected by different degrees of market power.[53]

The western United States is undoubtedly the part of the world where water markets have been most thoroughly studied. In this region—where water resources are scarcer than elsewhere in the country—water markets emerged as an important policy instrument and political issue in the 1970s and 1980s, when the United States entered a "new era" of water management. During the previous decades, national and state governments met growing demands for water by building dams and reservoirs and developing new water supplies. Since the 1970s, however, financial and environmental constraints have forced a change of emphasis to reallocation and protection of existing resources. Hence water policy debates in the United States have revolved around the same themes that more recently have come to dominate the international water policy arena: increased concern for economic efficiency and market incentives on the one hand, and increased concern for integrated water resources management, a basinwide focus, and environmental restoration on the other.[54]

Although many of the issues in contemporary international water policy debates are also prominent in the United States, most participants in the U.S. debates are unaware of the similarities, and the debates have proceeded in almost entirely domestic terms with little reference to international issues or organizations.[55] For example, the Dublin Principles are virtually unknown, and the World Bank and United Nations are not relevant actors. A more important contrast is that U.S. discussions and analyses of water markets have been more balanced and pragmatic and less ideological than has often been the case in the international and Latin American contexts. Economic perspectives have been broader, regulatory institutions have been stronger, and nongovernmental organizations have been active and informed participants. In consequence, it is generally understood in the United States that water markets are an important part of the larger water management context, but not the controlling principle. In Chile that view is not so widely shared.

CHAPTER 2

THE FREE-MARKET MODEL: CHILE'S 1981 WATER CODE

Chile's 1981 Water Code is a classic example of what in Latin America is often called the law of the pendulum: the historical tendency to swing from one extreme to the other in political and economic affairs, without finding a point of balance somewhere in the middle. Before 1967, Chilean water law was characterized by a relatively balanced combination of public regulation and private property rights. In 1967 the Agrarian Reform Law swung the pendulum toward greatly expanded governmental authority over water use and water management, at the expense of private rights; 14 years later the Chilean military government swung the pendulum to the opposite extreme, where it remains to this day.[1]

In general terms, the 1981 Water Code greatly strengthened private property rights, increased private autonomy in water use, and favored free markets in water rights to an unprecedented degree.[2] For the first time in Chilean history, the new Water Code separated water rights from landownership, created several market mechanisms and incentives, and aimed to foster a commercial and market-oriented economic mentality among water users (the vast majority of whom are farmers). As a corollary, the code sharply reduced the government's role in water resources management, regulation, and development. The Water Code's essential philosophy is laissez faire: it does not directly mandate or establish a market in water rights but instead aims to set up the legal rules and preconditions for such a market to emerge spontaneously, as a result of private initiative. In all these respects, the Water Code closely reflects the legal structure and ideological principles of Chile's 1980 Constitution, as discussed briefly below.

In more specific legal terms, the Water Code declares that water resources are public property and that the national government may grant private rights to use that property. These rights are called *derechos de aprovechamiento,* or "rights of advantageous use." Waters themselves are formally defined as *bienes nacionales de uso público,* "national property for public use," a term that dates to the 1850s, when Chile adopted its Civil Code, which is still in effect. It refers to a category of property that is owned by the nation as a whole and whose use belongs to all of the nation's inhabitants; other examples include roads, streets, bridges, plazas, and beaches. According to the definition in the Civil Code, such property cannot be alienated from public ownership or become the object of private legal or commercial transactions. The government can grant permits or concessions to private parties for the exclusive *use* of such property, but historically such use rights were governed by public (administrative) law and could be administratively modified or cancelled without compensation.[3]

Despite that formal legal definition, however, the Water Code in fact strengthens private ownership and control over rights to use public waters. Water rights are now legally separate from landownership and can be freely bought, sold, mortgaged, inherited, and transferred like any other real estate. There is a government water rights agency, the General Water Directorate (Dirección General de Aguas, or DGA), which is required by the code to grant all requests for new water rights, free of charge, whenever the water is physically and legally available. Once these rights are constituted, however, they are governed by private law rather than public law (i.e., by civil law rather than administrative law); this underlines their status as ordinary commercial goods and gives them significantly more protection from government regulation. Water rights are included in the general system of registration of real estate titles. Furthermore, water rights are now explicitly guaranteed as private property in the Constitution, as discussed below.[4]

Private liberties are very broad and government authority is tightly restricted, compared with earlier Chilean water legislation. Water rights owners can freely change the types or methods of use of their water rights without administrative approval from DGA (the one exception is changing the location of diversions from a natural channel). Applicants for new rights no longer have to specify or justify their intended water uses to DGA, and the agency has no discretionary power to deny such requests if water is available or decide among competing applicants. The code establishes no legal priorities among different types of water use, instead leaving such determinations to private parties and the market. If there is not enough water to satisfy simultaneous applications for new rights, in theory DGA must hold a public auction and sell the new rights to the highest bidder.

Other than such auctions, which in practice have been very rare, the owners of water rights do not pay taxes or fees of any kind, either for acquiring new rights from the government or for holding rights over time. Furthermore, owners have no legal obligation to actually use their water rights, and they face no penalty or cancellation for lack of use. The unconditional nature of private water rights differs from all previous legislation in Chile and also from the water laws of all other countries around the world.[5] (In the United States, for example, a legal doctrine called the beneficial use doctrine requires that private holders of water rights make some "beneficial use" of their rights or risk forfeiture; this is popularly known as the use-it-or-lose-it doctrine.) The 1981 Water Code omitted such legal and financial obligations because the code's drafters considered them undesirable restrictions on private property rights and economic freedom. All in all, these provisions allow unrestricted speculation in water rights, which has been one of the code's most controversial aspects, as discussed in Chapter 3.[6]

DGA now has very little regulatory authority over private water use. Nearly all decisions about water use and management are made by individual water rights owners or by private organizations of canal users—that is, irrigators. (These canal associations have a long tradition in Chile of building and operating canal systems and distributing water to their farmer members.) DGA cannot cancel or restrict water rights once they have been granted or otherwise constituted except by expropriating and paying for them, which is extremely rare.[7] The agency has also lost its power to adjudicate conflicts between water users; these conflicts now go to the ordinary civil courts. (Chile does not have any special administrative courts.)

DGA retains several important technical and administrative functions, such as gathering and maintaining hydrologic data; inspecting large waterworks, such as dams and canals; enforcing the rules governing the functioning of private water users' associations; and keeping official registries of certain water rights. (These registries are very incomplete, however: they include the rights originally granted by DGA but not the many rights constituted under earlier laws, nor do they record transactions after the original grant.) The agency can also prepare studies, reports, plans, and policy recommendations, but these have little or no regulatory force.

The Water Code's laissez faire principles are especially clear in the areas of river basin management and coordination of multiple water uses. Because the code's main concern was irrigation, it says very little about other water uses or about how to coordinate them. The exception was the creation of a new kind of property right, the nonconsumptive water right, which was the Water Code's major innovation in the area of multiple water use. Nonconsumptive rights were intended to stimulate hydroelectric development in the mountains and foothills upstream from agricultural areas, without injuring downstream farmers with preexisting water rights (now called consumptive rights). As the term implies, a nonconsumptive right allows its owner to divert water from a stream and use it, as long as the water is then returned unaltered to its original channel, for use by others downstream.

Beyond establishing the existence of nonconsumptive water rights, however, the legal rules governing their exercise and their relationship to other rights are very brief and general. As a result, coordinating different water uses depends on the Water Code's overall logic and institutional structure rather than on specific provisions: in other words, it depends on private bargaining among the owners of water rights rather than on government regulation, in a process favored by the Coase theorem.[8] These arrangements also reflect the Chilean Constitution, with its general model of strong private economic rights, limited state regulation, and a strong judiciary, as discussed below.

The Water Code recognizes several kinds of private water users' organizations, but all were designed solely for purposes of irrigation—to distribute water to canals and farms, not to settle conflicts with nonirrigators. Since DGA has no regulatory power to intervene in these matters, the tasks of coordinating multiple water uses and resolving river basin conflicts have been left to the free market—that is, to private bargaining among property owners. When this bargaining fails, the only recourse is to go to the ordinary civil courts, despite judges' lack of knowledge or experience of water issues. This framework has failed to prevent serious river basin conflicts, including those between the owners of consumptive and nonconsumptive rights, as discussed in Chapter 4.[9]

It is important to note that the Water Code does not address issues of environmental protection or water quality. Environmental issues were not a public or political concern in Chile in the 1970s or early 1980s, and the dangers of water pollution for public health were eventually addressed by other legislation. In Chapter 3, I briefly describe Chile's current environmental legislation, which was passed in 1994, in the context of the evolving debate about reforming the Water Code. In Chapter 4, I look at some of the problems that the Chilean government has faced since 1990 in trying to address environmental aspects of water management within the framework of the 1981 code.

THE 1980 CONSTITUTION: FOUNDATION OF THE WATER CODE

To understand the legal and institutional framework of the 1981 Water Code, it is crucial to understand that it reflects the structure of the current Chilean Constitution, which was adopted in 1980. This constitution was the creation of the military government and its civilian advisers. Like the Water Code, it was drafted in the late 1970s, without public discussion or opposition, at a time when the absolute political control of the military was accompanied by the ideological ascendance of the government's neoliberal economists. The 1980 Constitution was carried over essentially intact through the transition back to democratic government in 1990—opponents of the military regime in the mid 1980s had to accept this constitution as a condition for the regime to allow the return to democracy—and it is still in full force today.[10]

The 1980 Constitution is the most important legal and institutional legacy of the Chilean military government. This comprehensive and ambitious document includes economic and social principles as well as political principles. The objective was to consolidate and institutionalize the deep and radical changes imposed on Chilean society during more than 16 years of military rule.[11] The Constitution guarantees the basic legal framework for a free-market economic model by defining very broad private property rights and economic freedoms and tightly restricting the regulatory authority of government agencies and the national Congress. This framework is enforced by a judicial system that now has greater power to review and overrule administrative and legislative actions. The Constitution also guarantees a great deal of continuing political power both to the military and to other conservative and nonelected political forces.[12]

Neither the Constitution nor the economic model whose institutional foundations it cements can be amended today without the full agreement of those conservative political forces. There were repeated but failed attempts during the 1990s to forge such an agreement to further democratize the Constitution, and since the late 1990s the national political balance has moved further to the right. Any constitutional reforms approved in the near future will likely be marginal and not affect the economic and regulatory issues examined in this book.

That constitutional and political context must be kept in mind when analyzing the debates through the 1990s about reforming the Water Code, as examined in Chapter 3. Because water rights are explicitly protected by the Constitution's section on property rights, the definition of water rights cannot be altered except by a constitutional amendment or a legal interpretation that has broad political support. Furthermore, the increased power of the courts, which stems from the Constitution rather than from the Water Code, has major consequences for resolving water conflicts and the related institutional arrangements for integrated water resources management, as discussed in Chapter 4.

"LEGISLATIVE HISTORY" AND POLITICAL BACKGROUND

It is impossible to fully understand either the 1981 Water Code or subsequent water rights issues in Chile without some recent historical background. This is especially true because of the law of the pendulum—in other words, to understand the current Water Code, we must know what its drafters were reacting against. In the rest of this chapter, therefore, I will briefly summarize Chile's previous two water codes (passed in 1951 and 1967) and then focus on 1976 to 1981, the period of the "legislative history" of the 1981 code.

I put *legislative history* in quotation marks because the Water Code was written under an authoritarian military government that had closed the national Congress and other democratic political institutions several years earlier. The military government relied heavily on civilian advisers and solicited the opinions of civilian political allies and interest groups in many areas of legislation and public policy, and water rights were no exception. However, the government's internal discussions and decisionmaking processes were generally hidden from public debate at the time, and it is hard to reconstruct them. Hence my purpose here is to highlight the concerns of the people within the government who drafted the Water Code in the late 1970s, as well as the political and economic factors that shaped the code as it finally emerged in 1981. This analysis draws on interviews with many people involved in the process, as well as contemporary government documents, newspaper articles, and other publications.[13]

This legislative history is unfamiliar to all but a few Chilean experts, and it has more than merely academic relevance. In the first place, as we will see later in this book, some of the dominant issues and arguments in the debates about reforming the Water Code in the 1990s are nearly exact echoes of the earlier debates within the military government. Few people in Chile, and almost no one outside Chile, are aware of this. As a result, crucial aspects of the current debates seem to be exercises in futility. This is either unintended irony or astute political strategy (or perhaps both) on the part of the reform's opponents.

In the second place, the specific political decisions about legal rules and economic incentives that were made in 1981 have h[illegible]
the subsequent performance of Chilean water market[illegible]
relevant institutional arrangements. Most economic [illegible]
kets, however, have simply taken those political dec[illegible]
for granted, as discussed in Chapter 4. Hence a remin[illegible]
sions were made, and what alternatives existed, is [illegible]
encourage more institutional and interdisciplinary ec[illegible]

CHILE'S FIRST WATER CODE (1951): PRIVATE RIGHTS PLUS STRONG GOVERNMENT

Chile's first Water Code, enacted in 1951, had placed the pendulum in a central position, establishing a balanced combination of private rights and public regulation.[14] In most respects the 1951 code closely resembles contemporary water rights legislation in the western United States.[15] It systematized traditional Chilean rules and practices concerning water rights, which dated back decades and in some cases centuries, while increasing government intervention. This larger government role reflected the contemporary Latin American political and economic context, in which a strong and active state was considered necessary to foster national economic development. Several of the 1951 Water Code's main features continue to resonate in Chilean water policy debates.

Like today's Water Code, the 1951 code established a formal legal doctrine and administrative procedure for granting private rights to use waters that were publicly owned, and those use rights were treated and protected as private property rather than as administrative concessions or permits. Water rights granted were registered in the local offices of real estate title (*Conservador de Bienes Raíces*), which also recorded any subsequent changes of ownership. The 1951 code also declared the establishment of a centralized agency for water rights administration within the Ministry of Public Works. (In fact, the Dirección General de Aguas was not created until 1969; until then, DGA's functions were performed by other agencies in the same ministry, which housed the government's irrigation programs.)

In other respects the 1951 Water Code was quite different from today's code. It strengthened the government's administrative authority and imposed significant legal conditions on private rights.[16] Most importantly, DGA could cancel water rights if their owners did not use them for a period of five years. Applicants for new water rights had to specify the intended water use as well as describe the physical works that would be necessary. DGA granted such applicants provisional rights, which became definitive only upon later proof that the works had been completed and the water put to actual use. If there were competing applicants for new rights to the same waters, DGA followed a statutory order of preference among different uses: first drinking and domestic purposes, then irrigation, hydroelectricity, and other industrial purposes. Rights holders were not allowed to change the specific uses for which their rights had been granted; instead they had to return their rights to the government and request a new right for the new use. Finally, water rights were legally tied to landownership and could not be transferred separately. In short, the 1951 code made no effort to encourage free-market economic logic.

CHILE'S SECOND WATER CODE (1967): SWINGING LEFT TO CENTRALIZED CONTROL

Chile's second Water Code swung the pendulum toward greatly expanded government control and was the extreme against which the 1981 Water

Code would later react.[17] This second code was a by-product of the controversial Agrarian Reform Law of 1967, which was passed by the government of President Eduardo Frei Montalva and his Christian Democratic political party. The decade of the 1960s, of course, was the high point of centralized governments' attempts at social and economic reform in Latin America. The Chilean Agrarian Reform aimed to expropriate and redistribute large landholdings, with the twin purposes of expanding the class of small landowners and modernizing agricultural production.[18]

In a country whose agriculture depends on irrigation, land reform also requires redistributing water, and so the Agrarian Reform Law included a section on water rights that was later published separately as a new water code.[19] To achieve two main objectives—facilitating land redistribution and increasing the efficiency of agricultural water use—the code favored centralized government administration rather than private initiative. In consequence DGA was finally established in 1969, after having been legally authorized in the 1951 Water Code.

The 1967 Water Code, like its parent the Agrarian Reform Law, required an amendment to the property clause of the existing (1925) Constitution. The 1967 constitutional amendment expanded the scope of the so-called social function of property and thereby restricted the scope of private property rights. Although this certainly reflected the reformers' ideological views, the more immediate objective was to allow the government to expropriate private land while deferring the required compensation, by paying the owners with long-term government bonds instead of cash. (Since Chile experienced high inflation at the time, as through much of the 20th century, such bonds would lose most of their value within a few years.) The 1967 amendment also declared *all* of the nation's waters to be "national property for public use," including waters that had been considered private since the 1855 Civil Code: streams and lakes contained within a single landed property and, more importantly, waters flowing in "artificial" channels, or canals. In short, the amendment allowed the expropriation without compensation of all existing private water rights.[20]

Although water rights continued to be called "rights of advantageous use," they lost their legal status as property rights and reverted to merely administrative concessions, governed by administrative rather than civil law. They could not be privately bought, sold, traded, or separated from the land to which they had been assigned except with administrative approval by DGA (which was almost never granted). As a result, water rights were no longer registered as real estate titles, and there were no records of any subsequent transactions—which, after all, were illegal under this code. The lack of recordkeeping would lead to serious confusion and uncertainty about water rights titles by the late 1970s, uncertainty that to some extent persists to the present day.

Government regulatory powers were extensive under the 1967 Water Code. This code had the ambitious objective of redistributing water rights according to new technical "standards of rational and beneficial use." Government scientists and technicians would establish the amounts of water needed for different crops under different agronomic and geographic conditions. Water rights would then be allocated or reallocated to particular land parcels according to the local standards of use. This system of water rights was, of course, tied closely to particular patterns of agricultural land use and cropping, which were then in the process of being altered by government planning as part of the Agrarian Reform.

At the level of river basins, the government had authority to declare particular basins "areas of rationalization of water use," including nonagricultural water uses. Within these areas the government could reallocate water rights according to the technical standards of use and other planning criteria. In addition, DGA was given adjudicative power over water use conflicts, and the role of the courts was greatly reduced.

Such a technocratic model of water rights administration demanded a high degree of institutional capacity and resources and would have been very hard to implement even under favorable circumstances. In Chile, however, the years after 1967 were increasingly unstable, especially after the election in 1970 of President Salvador Allende and his Popular Unity coalition of leftwing political parties. The worsening social and political polarization, in the Agrarian Reform process as in other sectors of the economy, finally culminated in the military coup on 11 September 1973.[21]

WATER RIGHTS AFTER THE 1973 MILITARY COUP: CONFUSION AND NEGLECT

In the period after the coup, the military government put an end to land expropriation and reversed the Agrarian Reform. During the mid to late 1970s, the regime adopted a series of policies that "normalized" the agricultural sector, in line with the new national economic model of free markets and strong private property rights then being implemented. These policies included confirming and strengthening private titles to expropriated land, encouraging an agricultural land market, and reducing the government's role in agricultural production and commercialization. A relatively small proportion of the expropriated land was returned to its former owners. Instead, the regime seized the opportunity to modernize the agricultural sector by selling much of the state land to private buyers and by subdividing a significant amount into small parcels and transferring them to a select group of peasant farmers, known as *parceleros* (many of whom later resold their parcels in the land market).[22]

Throughout this process, however, the military government left water rights and the 1967 Water Code essentially untouched for more than five years: they were secondary priorities in the wider context. As a result, by the late 1970s the country's water rights situation was a mess. Chile continued to have a highly state-centered water law that was incompatible with the country's new economic model. The legal insecurity of private water rights discouraged private investment in water use or management, and the Water Code's rigidity prevented water transfers to more valuable uses. Water rights titles and transactions were especially uncertain because they had not been recorded since 1967.

NEOLIBERAL LEANINGS (1976–1981): PRIVATE PROPERTY AND FREE MARKETS

The preceding summary explains why water rights reform was necessary in Chile by the second half of the 1970s, but it was not at all clear what exactly that reform would look like. There was broad agreement within the military government and among its civilian advisers and supporters that water rights needed greater legal protection as private property. However, there was strong and sometimes bitter disagreement over whether such reform should be designed in the image of the free market and whether it required a drastic reduction in government regulation and spending.

The disagreements were at two levels: between different political and ideological viewpoints on one hand, and between different disciplinary and professional viewpoints on the other. In the first place, the conflict over water rights was only a small part of the more general political and ideological struggle within the military government between neoliberals (mostly economists) and more traditional conservatives. Many of the neoliberals were known as the Chicago Boys because of their graduate training in economics at the University of Chicago, which was known for its free-market viewpoint.[23]

In the second place, the conflict reflected the contrasting perspectives of economists, lawyers, and engineers. In Chile, engineers have historically been the professionals with the most knowledge and experience of water use and water rights (along with irrigators themselves, of course), followed by a small number of lawyers. Economists, in contrast, played essentially no role in water policy and management before the 1970s. Nevertheless, despite their lack of experience with water issues, the economists ended up having the predominant influence on the reform process and the resulting Water Code.

The Constitutional Commission (1976): Private Property But Not Free Markets

The more conservative position dominated the first discussions. In early 1976, the military regime's Commission for the Study of the New Constitution (also called the Constitutional Commission) took up the issue of water rights as part of its more general consideration of property rights.[24] Commission members, all civilian lawyers, had some initial doubts about whether material as technical as water rights should be given constitutional rank. Because of the water problems in the agricultural sector, however, they decided to include a brief statement of principle favoring private property as a first step toward greater legal security. They left the bulk of the 1967 Water Code in effect until it could be replaced by more systematic legislation.

Advised by prominent irrigation engineers, the members of the Constitutional Commission essentially advocated a return to the 1951 Water Code, combining private rights and government regulation. Their main argument was that the legal insecurity of water rights under the 1967 code had removed private incentives to build and maintain irrigation canals. These canals were said to be deteriorating throughout the country, and private associations of canal users were said to be in decline for the same reason. Strengthening water rights, the commission members believed, would stimulate private investment in irrigation works and revitalize the canal users' associations.

The commission members' position was far from promarket, however. Most of them emphasized the public aspects and obligations of water use, and they rejected a proposal to allow water rights to be sold separately from land as an incentive to increase water use efficiency. They concluded instead that such transactions should be restricted to certain specified situations and carefully regulated by legislation.

In the end, the Constitutional Commission agreed on the following statement, which was eventually included in the 1980 Constitution: "The rights of private parties over waters, when recognized or constituted according to law, will grant their owners property over those rights."[25] Note that it is property over the *rights*, not over the waters themselves, that is constitutionally protected, following the legal distinction mentioned in the summary of the 1981 Water Code at the beginning of this chapter. Note also that this statement protects rights dating from prior legislation. The commission deliberately rejected using the existing term *derecho de aprovechamiento* ("right of advantageous use") because of its connotations of public ownership and administration, although the term would later reappear in the 1981 Water Code.

However, when the military *junta* dictated its Constitutional Act 3 six months later, in September 1976, it omitted the commission's proposed statement and said only that "a special statute will regulate everything concerning mining property and ownership of waters."[26] Meanwhile, the 1967 Water Code remained in force. Why the more substantive statement was dropped is not entirely clear, but it seems that the members of the *junta* and their military staff had some doubts about the ambiguous nature of ownership of water—in particular about whether water could be owned by the government only or also by private individuals.

Less than two weeks before Constitutional Act 3 was published, the president of the Commission for the Study of the New Constitution, Enrique Ortúzar, and its most influential figure, Jaime Guzmán, met with the *junta* to advocate for their proposed statement. The *junta's* legislative secretary, a military lawyer, contended that waters should remain in public ownership. Ortúzar agreed with him in theory but said that the idea was to increase the legal security of use rights by treating them in effect as property rights. Apparently he was not persuasive: the commission's statement was shelved for almost three years.[27]

Decree Law 2,603 (1979): Swinging Right to the Free Market

A more neoliberal position became dominant in 1979, when the Chicago Boys had risen to control most aspects of the regime's economic and social policies.[28] In April 1979, the regime dictated the strongly promarket Decree Law 2,603, its first substantive legislation on water rights, which laid the foundation for the new Water Code two years later. Although this Decree Law repealed fundamental elements of the 1967 Water Code, it nonetheless left the bulk of that code intact until an entirely new law could be enacted.[29]

In addition to the strong ideological forces, two policy issues drove this legislation. First, by the late 1970s the need to clarify the nation's confusing water rights situation had become a priority. As noted above, the military government had completed its agrarian counterreform, the land that had been expropriated before the coup had now been sold or transferred to private owners, and the 1967 Water Code was incompatible with the increasingly active private land market. Many people argued that water rights had to be privatized to prevent future attempts at land reform: the major worry was that some future government might try to intervene in land tenure indirectly, through the control and reallocation of water rights.

Second, the government's neoliberal economic team was resisting pressure from agricultural interests to return to the government's historic role in irrigation development—repairing deteriorating dams and canals and building new projects that would make additional water supplies available.[30] The neoliberals had criticized these former irrigation policies as a classic example of the economic inefficiency of government action, and as a result the military government had sharply reduced spending on irrigation projects. In opposing a strong government role in irrigation, the neoliberals attributed the country's problems of water scarcity to the generally low level of irrigation efficiency and the predominance of low-value water uses. They blamed both phenomena on the state-centered and antimarket logic of the 1967 Water Code. As an alternative solution, they sought to create economic incentives for private investment in irrigation construction and maintenance.

Decree Law 2,603 strengthened private property rights to water by amending Constitutional Act 3 to read as the Constitutional Commission had originally suggested: "The rights of private parties over waters, when recognized or constituted according to law, will grant their owners property over those rights." (This provision was then repeated a year later in the new constitution.) The Decree Law separated water rights from landownership for the first time in Chilean history and allowed them to be freely bought and sold. It reestablished the system of registering water rights in real estate title offices and required that all water rights transactions be recorded in those registries as well. The new law did not explicitly make rights tradable, but it did so implicitly by requiring that trades be recorded. The Decree Law also attempted to "regularize" the uncertainty of existing water rights titles by declaring a presumption of ownership in favor of those who were currently using water rights de facto (Article 7) and by proposing to hold public auctions for all expired or cancelled rights.

A major innovation in the new law established that water rights were to be taxed like any other real estate. The idea was that water and land would be appraised and taxed separately, with the total not to exceed the taxes formerly paid on the irrigated land. Under the existing system, water rights were taxed only indirectly through land taxes, since irrigated land was much more valuable than nonirrigated land.

The government's economists argued that this set of policies would boost economic efficiency as well as water conservation by encouraging the owners of water rights to regard water as a commodity and an economic good rather than as a free (or at least unpriced) attribute of landownership. The economists further argued that the efficiency of water use would improve only if water had a real economic cost, reflected in higher prices, and if rights were defined as private, exclusive, and transferable. With higher water prices and the freedom to sell water rights separately from land, rights holders would have an economic incentive to invest in better irrigation technology and management, since they could then sell the rights to the water saved. The annual water rights tax would be an added incentive for rights holders to sell any unused or surplus waters and thereby reduce their tax burden.

Finally, although the law was directed primarily at the agricultural sector, its drafters also hoped to encourage intersectoral transfers. In other words, they aimed to improve irrigation efficiency so that the excess waters could be transferred to more highly valued uses, both within agriculture and in urban and industrial sectors, whose demand for water was increasing.[31]

The free-market logic of Decree Law 2,603 was fiercely debated within the government, and its provisions were somewhat diluted as a result. The original draft of the Decree Law had been even more promarket, with sweeping rhetoric about the benefits of fully commoditizing water rights and a statement that explicitly declared water rights to be fully alienable and allowed their uses to be freely changed. The draft proposed to allocate all available water rights by public auction—both existing rights that had been cancelled and future rights as well.[32] In a meeting with the *junta* in February 1979, two neoliberal cabinet members—Miguel Kast, Minister of

National Planning, and Alfonso Márquez de la Plata, Minister of Agriculture—cited the problems caused by the 1967 Water Code and defended the market logic of the proposed law. The two ministers emphasized that in addition to privatizing water rights, separate water rights taxes were the crucial mechanism of the new system; this would give water a real cost and create the incentive for its efficient use.[33]

But several military lawyers present were concerned about the public ownership of water, and they warned that neither the property status of use rights nor the validity of the term *derecho de aprovechamiento* was clear. Although the *junta* provisionally approved the draft law, subject to clarification of those doubts,[34] the version that was finally published two months later lacked the strong promarket language. The Decree Law dropped the sweeping declaration of free-market logic and the full commoditization of water rights and referred more modestly to the "national necessity of initiating the process of normalization of everything related to waters and their different forms of beneficial use," consistent with the regime's general economic principles. Instead of auctions of both cancelled and new water rights, the law established the former only.[35]

The neoliberal arguments were also opposed by most Chilean water experts, who were primarily engineers, lawyers, and irrigators rather than economists. Although their political views were generally conservative—many were civilians working in the military government—these experts' professional training and experience led them to place more importance on the various public aspects of water resources than did the economists.

The main political opposition, however, came from the agricultural sector. Irrigation, after all, was by far the country's biggest water use. With the Agrarian Reform still fresh in their minds, farmers and landowners were far more concerned about private property rights than about market efficiency or incentives. The most important agricultural interest groups supported the new Decree Law as part of the general process of national modernization, but they downplayed its market features. For example, in 1979 both the National Agriculture Society and the Confederation of Chilean Irrigators echoed the argument that had been made three years earlier by the Constitutional Commission—namely, that more secure private property rights would boost private investment in water use. Although in principle these organizations supported the freedom to change water uses and transfer water rights without government interference, in practice they expected few transactions and little reallocation.[36] This reflects the overall political position of these interest groups during the late 1970s and 1980s: politically loyal to the military government, they nevertheless protested the negative impact of neoliberal economic policies on agriculture.

Farmers' lack of interest in water markets was underlined when the new Water Code was published in October 1981. The president of the Confederation of Irrigators again praised the guarantee of security of property, but he doubted that there would be much trading of water rights. Market transactions, he said, "will probably not play an important role since the great majority of water rights have already been awarded. They might be relevant in more virgin rivers." He either did not understand or disagreed with the economists' objective of reallocating existing water resources. He was also uncertain about how market incentives might work in practice. Rather than investing in more efficient technology to sell the water saved, he thought that irrigators might sell water rights first in order to finance such investments.[37]

THE FINAL VERSION (1981): COMPROMISE AND PROBLEMS FOR FUTURE REFORM

The Water Code that was finally enacted in October 1981, more than two years after Decree Law 2,603 had laid the groundwork, was a compromise between the neoliberal economists and the less market-oriented conservatives. The neoliberals got most of what they wanted: a laissez faire legal framework that allowed private market transactions of water rights, and tight restrictions on government spending and regulation in the water sector.

Both the influence of the neoliberals and the close connection between water rights reform and government irrigation policy were shown by the new Water Code's companion statute, a law passed at the same time that established the legal and budgetary norms for financing government irrigation projects.[38] Those norms were so demanding that no new public irrigation projects were approved during the rest of the military government. (The government did, however, agree to subsidize small to medium-scale private irrigation projects in 1985, by passing a separate law at the request of the National Agriculture Society and the Confederation of Chilean Irrigators. This law was tacit admission that the Water Code's market incentives had not worked as hoped to stimulate private investment or water conservation, although the government glossed over that implication.[39])

Nonetheless, the neoliberals had to give up the legal rules and financial provisions that would have raised water's cost and price—measures that they had argued were crucial incentives for market discipline and efficiency. The 1981 code abandoned the proposed system of water rights taxes and did not impose any other fees, whether for acquiring new water rights from the government or for holding rights over time. In addition, the code required public auctions for a significantly smaller category of water rights than had been proposed earlier: auctions would not be held for all new rights or for cancelled rights, but only when there were two or more simultaneous requests for the same new rights. These auctions have proven to be quite rare.

The establishment of some kind of tax or fee was blocked primarily by the agricultural sector, which as the nation's predominant water user was the most affected by the new water law. Most farmers were in financial trouble after years of painful adjustment to the free-market economic model, and they could not afford (or simply refused) to start paying for a resource that had always been free. They were evidently not convinced by the neoliberals' argument that the total amount of land and water taxes would not rise once the payments were separated.

There were also practical and administrative objections to a system of taxes or fees. Given Chile's highly diverse geography and the legal and technical uncertainty of many water rights titles, the task of determining the appropriate rates for water rights owners throughout the country and enforcing their collection would be both technically and administratively difficult and politically unpopular.[40]

Hence the neoliberals had to settle for a market-oriented scheme that depended on much weaker price signals and incentives than they had argued for. Although they would have preferred a system of water rights taxes to change water users' economic thinking and induce water rights trading, they hoped that with a permissive legal framework there would be enough voluntary transactions that a market would eventually take shape. The neoliberals were less concerned about losing the fees for granting new rights, since their primary objective was to limit administrative discretion and transfer ownership from public to private hands. Twenty years later, these issues and arguments are still at the heart of Chilean water policy debates, as we will see in the next chapter.

In conclusion, the 1981 Water Code reflected the deeper political tension within the military government between neoliberals and an assortment of other conservatives. The two sides generally agreed on strengthening private property rights and weakening government regulation but clashed over the full application of free-market logic. The disagreement highlighted their different conceptions of the nature and purposes of private property. The neoliberals saw property rights as commodities, the basis for private bargaining and market transactions that would lead to increased economic efficiency. Their conservative opponents, on the other hand, including the great majority of irrigators and water users themselves, wanted private autonomy and legal security as ends in themselves but were much less interested in the full commoditization of water rights, and they did not share the objective of ending government subsidies. This fundamental distinction is often lost in political and policy debates, both in Chile and elsewhere, in which agricultural interests' defense of property rights is often mistaken for a defense of the free market. Many people blur this distinction deliberately, of course, for reasons of political strategy.

The history of the Water Code also demonstrates one of the basic premises of institutional economics, as discussed in Chapter 1: rules and institutions come before markets, not the other way around. Political decisions and legal rules about property rights have a major impact on price signals and economic incentives, and therefore on how—or whether—particular markets work. Thus institutional arrangements shape the way in which markets determine value and allocate resources. Because all markets are built on such prior political decisions, formally expressed as legal rules and institutions, even a "free" market cannot be neutral, objective, or apolitical, as proponents often claim.

The Chilean military government's decision not to impose taxes or fees on water rights is easy to understand, given the political and economic context, and the decision can be readily justified. Nonetheless, that decision then became cemented constitutionally, and those legal and economic conditions became vested interests and rights. In consequence, any subsequent imposition of a different legal rule or economic incentive on existing water rights must overcome high constitutional and political barriers, as discussed in the next chapter.

Finally, I want to underline two points that I made earlier this chapter, in summarizing the Water Code. Both points have proven very important since 1990, as we will see in the following chapters.

First, the creation of the new category of nonconsumptive water rights—to encourage hydroelectric development without affecting farmers or other water users—was basically an afterthought. The idea was not mentioned before 1979, and it simply appeared when the new Water Code was published in 1981. The lack of debate about the idea, even within the closed circles of the military government and its civilian supporters, helps explain why the legal rules defining these rights were so vague, with such lasting negative consequences after 1990.

Second, because agricultural water use was the dominant priority and concern among those who wrote the Water Code, the legal and institutional arrangements for other water management issues were either overlooked or simply left to the free market. Problems of river basin management, coordination of multiple water uses, conflict resolution, economic and environmental externalities, and so forth would have to be handled by the general framework of the 1980 Constitution, which established strong private property rights and economic freedom, weak government regulatory agencies, and a powerful but incompetent judicial system.

CHAPTER 3

REFORMING THE REFORM? POLICY DEBATE UNDER CHILEAN DEMOCRACY

If we pause in early 2004 to review the Water Code's first 20-plus years, what trends stand out? I will briefly highlight three general points and then discuss the second and third points in detail in the next two chapters.

First, the Water Code's second decade (the 1990s) has been much more dynamic and controversial than its first decade (the 1980s). During the 1980s water rights issues had a very low public profile in Chile and provoked little public debate. One reason is simply that the Water Code was still new and fairly unknown, even among water users. Another is the general weakness of the national economy until the later 1980s: the poor economic conditions reduced demand for water resources and hence lessened the impact of the new law.

The major reason for the lack of interest in water issues during the 1980s, however, was the wider national political context. The military government was in power throughout the decade and maintained strict control over political debate and public policymaking. Water law was a very minor concern in the context of the long and difficult struggles between the regime and its various opponents about when and how to make the transition back to democratic government. The national plebiscite in October 1988, in which General Pinochet's bid to rule the country for eight more years was rejected, followed by national elections in December 1989 to choose a new president and Congress, dominated public attention. It was only in 1990 that political conditions really opened up and more prosaic and technical policy issues could return to public debate.[1]

It was the Water Code's second decade, therefore, that marked its true emergence onto the public stage and into public awareness in Chile. The Water Code's profile was further raised by the country's rapid economic growth after 1986, which increased and diversified pressures on water resources. (In the 10 years after 1986, the gross national product grew at an average of about 7% per year; since then the growth rate has dropped by about half but has nonetheless remained steadily positive.) For both political and economic reasons, then, I will focus my analysis in the next two chapters on the period since 1990.

The second general point that stands out in retrospect is that virtually the entire decade of the 1990s has been characterized by strong political disagreement in Chile about whether to reform the Water Code, and if so, how. Since 1990 three successive national governments have proposed a series of legal and policy reforms in the areas of water rights and water management. (All three governments have been formed by the same coalition of center-left political parties, known as the Concertación, that has governed Chile since the return to democracy in 1990.) These proposals have generated heated political and professional debate, and so far they have been blocked by opposition from right-wing political parties and private sector interest groups. As of early 2004, after more than 10 years of debate, the fate of the reforms remains uncertain. What is clear, however, is that any measures approved in the near future will be much more limited than the reformers had hoped in the early to mid 1990s.

The third point is that research on the Water Code's results has led to a gradual improvement in the level of empirical knowledge and understanding. Foreigners have done as much of this research as Chileans have, in both cases including academics as well as nonacademics. Like the political debates mentioned above, this research began in the early 1990s, and nearly all the attention has focused on water markets and water rights trading.

Within Chile, the empirical research has been closely related to the policy debates about reforming the Water Code. These debates have been highly politicized and ideological, which inevitably has affected the empirical arguments and analyses—a factor that outsiders to Chile have typically not understood. Nonetheless, the gradual increase in empirical knowledge has meant that by the late 1990s, consensus was emerging about how to describe Chilean water markets in practice. Political disagreements about how to interpret that description have continued to be strong.

In the next two chapters I examine the evolution of those two major themes through the 1990s and into the 21st century. In this chapter I discuss the policy and political debates within Chile about reforming the Water Code; in Chapter 4, I turn to the academic and empirical analyses of the Water Code's practical results. Issues of water markets and property rights have dominated people's attention in both arenas. I will argue, however, that it is also important to emphasize the issues that have been either overlooked or looked at only superficially—namely, social equity, river basin management, conflict resolution, and environmental protection. The slighting of these issues in Chile reflects the constraints of national politics as well as the intellectual and cultural predominance of a narrow and free-market perspective on economics. Outside Chile, however, these issues are not at all secondary: they lie at the heart of recent international water policy debates, as discussed in Chapter 1.

REFORMING THE WATER CODE: *MUCHO RUIDO, POCAS NUECES*[2]

After more than 16 years of military rule, Chile returned to democratic government in March 1990, following national elections in December 1989. The Concertación candidate, Patricio Aylwin, won the presidency, and the coalition also won a majority in the Chamber of Deputies. The new government's major objectives were to consolidate the transition to democracy, maintain market-oriented economic growth based on exports of natural resources, and reduce poverty and social inequity.

The political and economic constraints on the new government were considerable because the military agreed to leave power only after the Concertación had firmly committed itself to respecting and working within the military government's institutional legacy: the 1980 Constitution, its associated legal and political framework, and the neoliberal economic model. Thus the new government had limited room to move in implementing its general program. Any significant legislative change required the consent of at least part of the political right, which enjoyed disproportionate influence in the national Congress, especially the Senate, and in the judicial branch (disproportionate, that is, in relation to the popular vote). The Concertación government has been reelected twice since 1989—Eduardo Frei, Jr., was elected president in 1993 and Ricardo Lagos followed him in 1999—and the overall political and constitutional context has remained essentially the same.[3]

Water law and policy have been secondary priorities in that broader context. Nevertheless, the Concertación developed plans to reform the 1981 Water Code in its first months in office, and both subsequent administrations have continued these efforts.

Throughout the 1990s and into the new century, the government's essential criticism of the Water Code has been that the law's neoliberal features are too extreme—that the pendulum has swung too far toward free markets and deregulation, and that it is time to return to a more balanced position. Hence the government has proposed legal changes that would alter important aspects of the Water Code but without swinging the pendulum back to the other extreme, as represented by the 1967 Water Code. From the perspective of the current Water Code's defenders, however, the government's approach has been either misguided or untrustworthy, and any movement of the pendulum toward a more central position would be a potentially dangerous increase in state control.

The core issue of Chilean water policy debates in the 1990s, therefore, has been profound disagreement about some of the most basic rules defining property rights to water, and hence also disagreement about the proper role of government and the proper scope of public regulations in water management. All the specific and technical issues in dispute reflect this overarching theme. Furthermore, the debates have taken place on at least two levels, and the arguments have often gotten mixed up. On one hand, people have argued about the legitimacy and fairness of the basic legal rules and definitions; on the other hand, they have argued about the rules' economic efficiency and practical impacts.

Over the course of the 1990s, there have been two major rounds of debate about reforming the Water Code, corresponding to two packages of proposals sent by the government's executive branch to Congress. The first package of proposed reforms, in the early 1990s, was aggressive and clumsy, and it was shot down within a year. The second package was more limited, careful, and pragmatic, and seven years later it is still being debated.[4]

ROUND 1 (1990–1993): THE GOVERNMENT GOES TOO FAR

The government's initial position in the early 1990s included a strong political and ideological critique of the Water Code. In 1990 the Dirección General de Aguas (DGA) was instructed to prepare a new "National Water Policy," in consultation with other government agencies, private sector interest groups, and nongovernmental organizations. Its charge was to diagnose the nation's water rights situation, identify the problems for which the current laissez faire approach was inadequate (such as environmental protection and other public interests), and propose the legislative changes needed to address those problems.

At the same time the government also named a prominent water lawyer, Gustavo Manríquez, to be the new head of DGA. Since all previous heads of the agency had been engineers, the nomination of a lawyer reflected the priority that the new government placed on reforming the law. Manríquez had participated in the military government's drafting of the Water Code ten years earlier, a fact that cut both ways: it meant that he was intimately familiar with the issues, and he evidently shared some of the new democratic government's criticisms of the code,[5] but it also raised questions about his own political and ideological views.

Early versions of the National Water Policy emphasized the importance of the "general interest of the nation" and argued in favor of "recuperating" water's status as a *bien nacional de uso público* ("national good for public use").[6] Although the 1981 Water Code defined water resources as having this status, and hence as public property, the government's view was that in reality the code had undermined the formal definition by fully privatizing the rights to *use* water.[7] In a series of public meetings and seminars, DGA representatives suggested a variety of legal changes, many of which would mean increasing restrictions on private rights and strengthening the agency's regulatory powers and duties. DGA eventually moderated some of its initial proposals in response to comments received in those meetings.

After two years of this process, the government sent its first round of proposed legislative reforms to Congress in December 1992. In this bill the government argued that the Water Code "suffers from excessive permissiveness and passivity" in the face of critical problems of water scarcity and contamination.[8] By defining private water rights in a manner that includes no duties or obligations to public interests, the code had favored private speculation, hoarding, and monopoly of water rights, and had undermined the incentives for using water rights in economically productive activities.

According to the government, this state of affairs was both socially inequitable and economically inefficient: inequitable because it allowed private interests to profit from public property without performing a useful social function in return, and inefficient because it held back national economic development by discouraging the productive use of resources.

The 1992 bill had four parts. The first part expressed the goal of "recuperating" water's public characteristics by altering the legal definition of property rights to water. The second part increased DGA's regulatory authority to protect water quality and watercourses, including the power to require future recipients of water rights to maintain minimum flows for ecological purposes. The third part addressed the existing institutional void in river basin management by proposing the creation of new basinwide organizations to administer water use. The fourth part made adjustments for different regional conditions—tightening regulations in the arid North, where water scarcity was a serious problem, and loosening them in the rainy South.[9]

The bill would have redefined the scope of private property rights in several ways. The most controversial was a proposal to return to the traditional legal rule that obtaining and owning water rights required putting them to concrete use. Thus an application to DGA for new water rights would have to specify the planned uses, and any rights not used for a period of five years could be cancelled by DGA without compensation and reallocated to other water users with more immediate and concrete needs. This use-it-or-lose-it rule had been part of Chile's two previous water codes and was familiar to Chilean water experts and water users.[10]

The government recognized that this change would affect water users' property rights and made two constitutional arguments in its favor. One argument was that even under the current Constitution, with its strong protections of private property rights, such rights were constrained by the "social function of property." The Constitution defines this social function to include "the general interests of the Nation, national security, public utility and health, and the conservation of environmental patrimony."[11] As the government argued, this definition was broad enough to allow some limitations on water rights.

The other, related argument was that the government could impose some conditions on private rights to use public property, since no one disputed that water resources themselves were still inalienably public property. The government went on to argue that the additional conditions and constraints would not in fact weaken the legal security of private rights, although they might appear to do so; on the contrary, in a larger sense the reform would actually *increase* that security by correcting the unregulated disorder of the water rights system as a whole.[12]

Although the government's general diagnosis of the water rights situation was reasonable, its constitutional arguments were debatable and its political tactics were clumsy and heavy-handed. The proposed use-it-or-lose-it rule became a lightning rod for opposition, generating fierce resistance from private sector interest groups (particularly those representing agricultural water users), from neoliberal economists, and from right-wing politicians. Some of these opponents shared the government's criticisms of speculation, potential monopoly power, and the lack of use of water rights, agreeing that these problems were obstacles to economic development and should be reformed. However, even these more sympathetic opponents rejected the government's approach to solving the problem.[13]

In the first place, nearly all of the government's opponents rejected the use-it-or-lose-it rule for being clearly unconstitutional: the government could not place new restrictions on vested property rights without paying compensation. This was certainly the intent of the military government and its Constitutional Commission in the 1970s, as described in the historical background of the Water Code (Chapter 2). Even if the Concertación government's legal arguments to the contrary might be theoretically plausible, they sounded too much like the Agrarian Reform and other interventionist policies from 20 years earlier.

This particular constitutional issue indicates the broader political problem: the government's opponents attacked the whole package of reforms for being an excessively centralized, "statist" (*estatista*) approach that was dangerously reminiscent of the conflict leading up to the military coup. From this viewpoint, the proposed reforms suggested that the Concertación's proclaimed commitment to the neoliberal economic model was shallow or insincere. The reforms would increase DGA's administrative discretion and hence the potential abuse of bureaucratic power, and they would undermine the economic model's fundamental elements: private economic liberty, security of private property, and the apolitical neutrality of the free market. In this context the use-it-or-lose-it rule was a symbolic example of the government's alleged antimarket position.

The proposed new river basin organizations were also controversial. According to the government's proposal, these organizations would be public-private hybrids at the regional scale. They were referred to as "administrative corporations," and they were supposed to bring together all the major public and private actors in basinwide water use, including water rights owners and canal users' associations, government agencies and public enterprises, municipal governments, local universities, private sector associations, and other nongovernmental organizations. Unfortunately, the government said nothing concrete about how the new organizations would work, leaving out such critical questions as how decisions would be made and enforced, what the proportions and voting rights of different members would be, where the financing would come from, what powers and duties the organizations would have in relation to existing government agencies and users' organizations, and so forth.[14] The government's purpose was presumably to initiate public discussion of these issues rather than to dictate the outcome, but the proposal was widely criticized as vague, heavy-handed, and further evidence of the government's antimarket intentions.[15]

The more extreme of the antireformers argued that the Water Code and the water market had worked very well and no changes of any kind were needed. They rejected virtually any additional government authority in any area of water management, even if evidently constitutional; for example, they opposed giving DGA the power to impose minimum ecological flow requirements on future grants of water rights, although no vested rights were affected.

Most of the government's opponents agreed that the Water Code had some flaws but insisted that the problems should be corrected in a manner compatible with the law's core market principles—that is, by adopting some kind of economic incentive instead of a legal requirement. The use-it-or-lose-it rule was criticized in economic terms for being a direct incentive for inefficient water use and allocation (as indeed it is commonly criticized in the western United States). A better alternative would be some form of tax, fee, or other economic instrument.

ALTERNATIVE INSTRUMENTS: WATER RIGHTS TAXES VERSUS FEES FOR NONUSE

Two alternative economic instruments have dominated subsequent debate in Chile, from 1993 to the present day. One is to establish annual taxes on water rights, just as on land and other real estate, which would be payable whether or not the rights were used. This, of course, was exactly the system advocated by the neoliberals in the late 1970s but rejected by the military *junta* because the political and administrative costs were too high (see

Chapter 2). I will return to this alternative later this chapter, because after languishing in the background for most of the 1990s, it has recently returned to center stage with renewed political force.[16]

The other alternative would establish an annual fee to be paid by the owners of any water rights that were *not* being used. This fee for nonuse, known as a *patente,* is a concept borrowed from Chilean mining law, which resembles Chilean water law in some essential respects. Mineral resources are defined in the Constitution as the inalienable property of the national government (though not as the property of the general public, as in the case of water). Private parties, however, can acquire permits to use and extract those minerals, and such permits, once granted, become property rights. If the owner of a mining right is using that right, he owes the government nothing; if he is not using it, however, he must either return it to the government or pay an annual fee (*patente*) to keep it. In this way the *patente* provides an economic incentive to use the resource.[17]

The government was open to both suggestions. In fact, DGA's early proposals in 1991 had mentioned both *patentes* and property taxes as alternatives worth exploring.[18] In 1993, a few months after the proposed bill had been sent to Congress, Manríquez, the head of DGA, told a congressional committee that some form of tax or fee had been left out of the bill "only with the purpose of letting it flow from the discussion itself."[19]

Six months later the government amended the bill to abandon the use-it-or-lose-it rule and replaced it with fees for nonuse. The amendment said that these fees would apply not to existing water rights but only to newly granted rights that had not yet been used by their owners; this was a fairly limited category, but one that raised fewer alarms about constitutionality. The government proposed fees that varied by region (much higher in the North, where water was most scarce) and increased over time if the rights remained unused. In addition, fees would be much higher for nonconsumptive water rights than for consumptive rights.[20] Nonconsumptive water rights would soon become one of the government's central preoccupations, as we will see below.

Despite the government's abandonment of the use-it-or-lose-it rule, by late 1993 the strength of the opposition to the proposed reforms forced the government to withdraw the entire bill from active discussion in Congress. National elections in December 1993 were another reason to pause and regroup. The Concertación government was reelected, as expected, with Eduardo Frei Ruíz-Tagle winning the presidency and the Concertación retaining the majority in the Chamber of Deputies, although not in the

Senate. Nonetheless, the turnover of pivotal politicians and administrators put the Water Code reforms temporarily on hold. One important change was at the head of DGA, where Manríquez was replaced by a senior DGA engineer, Humberto Peña. Peña was still in this position in 2003, providing valuable continuity of leadership and overseeing a long-term process of administrative modernization of the agency.

GROWING EMPHASIS ON ECONOMIC INSTRUMENTS AND ANALYSIS (1994–1995)

After the failure of the first round of reforms, the debate about changing the rules and incentives for water use continued steadily for the next several years, although with a lower public profile. The discussion became more technically sophisticated as the government developed and refined its arguments in preparation for another attempt at legislative reform and also made strategic and tactical adjustments to gain political support. This period marked the beginning of a notable increase in economic analysis and commentary, among both policymakers and academics. The attention of economists—primarily foreign academics rather than Chileans—was focused on the performance and results of Chilean water markets.

Although the government had abandoned the use-it-or-lose-it rule and embraced the concept of a more directly economic instrument, there were different views about how to proceed. The predominant view within the government favored the establishment of *patentes,* fees for nonuse that would apply at least to all newly granted and still unused water rights; some people in the government wanted to apply them to unused existing rights as well. The preference for fees for nonuse reflected the belief that such fees would be a good first step in reform: easier to implement and less politically sensitive than imposing new property taxes on all new and existing water rights.

For example, the Minister of Public Works, Ricardo Lagos, explained in 1994 that the government's position was that the fees for nonuse were "perhaps a minimum element" in the larger context of water problems. He was optimistic that Congress would approve the fees quickly so that they could all turn their attention to more difficult water policy issues, such as long-term water planning, river basin coordination, and environmental management. Those more complex issues would require more extended analysis and negotiation.[21] (The same Ricardo Lagos has been president of

Chile since 2000, and he was already a prominent political figure when he was minister of Public Works.)

Others within the government had more ambitious ideas. The head of the Irrigation Directorate (DGA's sister agency within the Ministry of Public Works) argued that fees for nonuse were an inadequate solution to the problem of speculation and lack of use of water rights. Such fees would affect only a small proportion of the country's water rights, he said, and they ran counter to the market logic of the Water Code. Instead, he recommended a more comprehensive approach to improving economic incentives and increasing water's economic value: establishing an annual fee (*tarifa*) for all uses of water as a "productive input" in economic activities. This fee would include a fixed charge plus a charge that would vary by the volume of water used. His reasoning was very similar to the neoliberal economists' arguments in favor of separate water rights taxes back in the late 1970s, and like those economists, he did not address the institutional feasibility of actually implementing such a system.[22]

AN ASIDE: CHILEAN ENVIRONMENTAL LAW

Chilean environmental law has so far had little effect on water rights or water resources management because modern environmental legislation is still quite recent in Chile and regulation is fairly weak. Environmental issues were not a priority under Chile's military government, but they were on the political agenda when the Concertación took office in 1990. Much of the pressure to pass a new environmental law was external. The Concertación wanted to strengthen Chile's international commercial relations both to increase foreign investment in Chile and to expand markets for Chilean exports, but beginning in the 1990s, international trade agreements required countries to have at least the appearance of effective environmental legislation.

Chile's first modern and systematic environmental legislation was passed in 1994, at the end of the administration of President Aylwin, and it was shaped by several years of political negotiation between the government, the right-wing opposition, and powerful business interest groups. The law established basic principles and a general framework.[23] It was not a comprehensive statute, but rather it laid the groundwork for more detailed and technical laws in specific areas, which were to be written later. The government's objectives were explicit and pragmatic: to proceed gradually; to inventory and organize the many existing legal norms, scattered among diverse sectoral laws and regulations, that concerned the environment; and above all, not to slow down national economic growth.

Those objectives have been met. The 1994 law established a new agency, the National Environment Commission, whose role is to "coordinate" other government agencies and ministries. The commission itself is both legally and politically weak. It has limited regulatory and enforcement powers and is directly subordinate to the office of the president, which has kept the commission on a tight leash. Environmental impact statements for specific projects were the substantive area that was most developed in the 1994 law and that have gotten most attention in subsequent regulations. Most other aspects of environmental protection and management have advanced more slowly.[24]

The pros and cons of Chilean environmental law and policy have been much debated in Chile, and this is not the place to discuss the issues in detail. For the purposes of this book, the important point is that matters of water rights, water markets, and water management have been essentially separate from environmental regulation. In their autonomy from the National Environment Commission, the Water Code and DGA have been like other sectoral laws and government agencies. This does not mean that DGA has ignored environmental issues since 1990—on the contrary. But for the most part, DGA has addressed these issues within the framework of the Water Code rather than the 1994 environmental law.

ROUND 2 (1996–2003): THE GOVERNMENT MODERATES ITS POSITION

The government's executive branch sent a second round of proposed Water Code reforms to Congress in July 1996.[25] The new proposals shared the overall diagnosis and many of the more technical provisions of the 1992 bill—that is, the government still argued that the code was too laissez faire to handle the country's growing water problems of scarcity, conflict, and environmental degradation. Nonetheless, the 1996 proposals differed in important ways.

Overall, the new proposals were more limited, pragmatic, and carefully prepared. In part this reflected the hard political experience of the first round of debate. But it is also clear that this change was more than simply a tactical shift: the government's policy analyses and positions in water issues became increasingly balanced, thoughtful, and consensus-seeking. Perhaps the best evidence of this change is the several years that DGA spent refining the documents describing the National Water Policy, which the agency then disseminated for public education and discussion in the late 1990s.[26] Another illustration of the agency's less confrontational attitude during this period is its decision to commission reports on technical policy issues from academic experts and institutions not generally supportive of the government's position, most notably the Catholic University in Santiago.[27]

A second change was the government's proposal to alter the basic rules defining property rights by establishing fees for nonuse that would apply to all existing and future water rights, as described above. The government attempted to alleviate the opposition's fears about too much administrative discretion in two ways: first by specifying the guidelines that DGA would follow in determining which rights (or which portions of rights) were not being used, and second by establishing judicial review of the procedures. Like the previous reform package, the new proposals also aimed to increase DGA's authority to impose conditions on new water rights to protect various public interests. For instance, to prevent speculation, the agency would require an explanation of the concrete need for the water as part of the application process for new rights, as well as a study of the interactions between surface water and groundwater. Furthermore, in granting new rights, DGA would have the power to consider and require the maintenance of "ecological flows"—that is, minimum in-stream flows for environmental and water quality protection.

Although those proposals aimed to increase restrictions on private ownership of water rights, a third change was that the government now explicitly declared its support for water markets, under the appropriate conditions. The government expressed its argument in conventional neoclassical economic terms: the Water Code's libertarian definition of property rights, and its failure to impose any costs on water rights ownership, had distorted the price signals about water's economic value and thereby undermined the efficient functioning of market mechanisms. The reforms, in contrast, were designed to correct those distortions and thereby improve the market's performance.[28]

As Public Works Minister Lagos explained in a 1995 newspaper interview, "if it is a question of making the market operate freely, then it is necessary to introduce the price of water as an input. This can be done either by charging a fee [for nonuse] or a fee proportional to the use which is made of the resource."[29] DGA's National Water Policy later elaborated the same argument: "Water is an economic good and as such the legal and economic system which regulates its use must encourage its efficient use by private individuals and by society as a whole. Accordingly, the principles of market economics are applicable to water resources, with the adaptations and corrections demanded by the particularities of hydrological processes."[30]

A fourth important change was that nonconsumptive water rights became an increasingly prominent and controversial issue. As described in Chapter 2, the creation and definition of nonconsumptive rights was one of the major innovations of the 1981 Water Code, intended to foster hydroelectric power development without harming the consumptive water rights of irrigators. By the mid 1990s nonconsumptive rights had become the principal target of the government's criticisms of speculation, hoarding, and failure to use existing water rights. This reflected problems of electricity law and regulation as much as problems of water law: the government was increasingly concerned about the monopoly power of private corporations in the nation's electricity sector, which the military government had privatized in the 1980s. Because electricity generation in Chile depends heavily on hydroelectric dams, issues of water policy and electric policy are closely related.

From 1995 to 1997, DGA was embroiled in a high-stakes legal dispute with the country's largest and most powerful private electric company, ENDESA, over the company's applications for new nonconsumptive water rights for future dam projects. DGA refused to grant the new rights on the grounds that such an accumulation of rights would severely damage competition. The conflict eventually went to Chile's national Antimonopoly Commission, which ruled in favor of DGA in early 1997.[31] The broader issues of the relationship between Chilean water and electricity policy, however, continue to be complex and difficult.[32]

In any case, the government's increased focus on nonconsumptive water rights was limited to the issues of fees for nonuse and concentration of ownership; it did not address the problems of river basin conflicts over water use or the tense relationship between consumptive and nonconsumptive water rights. This was a remarkable omission, given that the previous several years had seen repeated legal clashes between farmers, power companies, and environmental groups over how to coordinate the two kinds of rights within a shared river basin. In the course of these conflicts, the erratic and superficial rulings of Chile's Supreme Court had shown the need to clarify the relevant provisions of the Water Code (see Chapter 4).[33]

Finally, another significant change in the second package of reforms was the indefinite shelving of the proposal to create new river basin organizations. The decision to drop that proposal was presumably related to the government's failure to address the competition between consumptive and nonconsumptive water rights. The need for better river basin management has continued to draw rhetorical attention in Chile and has been the subject of a steady trickle of internationally funded technical studies. As a real political possibility, however, the issue is off the table.

All of those changes reflected the government's political assessment that the next step in reforming the Water Code should be to adopt the measures about which there was broad consensus, while leaving the more conflictive issues for later discussion. This was expressed in the president's message to Congress presenting the bill,[34] among other occasions.

THE OPPOSITION HARDENS

Despite the government's efforts to respond to earlier objections and to present a more moderate package of reforms, the new proposals met strong opposition from the same set of private sector interest groups, neoliberal policy analysts, and right-wing politicians. Again the opposition focused on the attempt to modify the basic rules defining property rights, this time the proposal to impose fees for nonuse of water rights. Some opponents initially found the general concept of these fees acceptable—after all, many of them had suggested or supported the idea several years earlier, as an economic incentive compatible with market principles[35]—but not the specific version proposed by the government. Private hydroelectric and mining interests, for example, complained that the fees were too high and the period allowed for nonuse was too short, particularly for major investment projects that required a longer time horizon. But they did not present any counterproposals or attempt to negotiate a compromise.[36]

By the end of 1996 the reform's opponents had dropped any earlier suggestion that they might agree to some kind of fee for nonuse. Instead, they criticized the measure as a perverse incentive that would encourage inefficient water use. As in 1993, the core of the opposition to reform was a deep political and ideological distrust of any increase of government regulation or any new restrictions on property rights. The antireformers accused the government of trying to expand DGA's discretionary authority and undermine the Water Code's free-market principles. The more charitable opponents thought the government was simply misguided; the less charitable saw the renewed threat of socialism, centralized planning, and the Agrarian Reform.

One prominent critic was a water lawyer who as a high-ranking civilian adviser in the Ministry of Agriculture during the military regime had helped write the 1981 Water Code. In an editorial in the conservative newspaper *El Mercurio*, he warned of "planning on the attack ... causing injuries to the freedom of enterprise."[37] The president of the National Mining Society, a private sector interest group, spoke in similarly incendiary terms: "The wide discretionary faculties that the bill gives to political officials ... will convert the Director of the DGA into *one of the people with the most economic power in the country: a true 'lord of water'* [*'señor de las aguas'*]."[38] The degree of rhetorical stridency varied, but other opponents of the reform argued in similar political and ideological terms.[39] In the face of such reactions, the government's willingness to negotiate such details as the fee structure and guidelines for implementation was to no avail.

Although the highly ideological tone of many the reform's opponents was undeniable, it is also true that some of their criticisms were reasonable and well taken, as discussed below.

CONSTITUTIONAL CHALLENGES OF FEES FOR NONUSE

In August 1997, after more than a year of unhurried discussion of the government's proposed legislation, the Chamber of Deputies voted in its favor and sent it on to the Senate. (At the time the Concertación had a majority in the Chamber, but in the Senate the balance with the right-wing political opposition was essentially equal.) Having failed to block the bill, a group of right-wing deputies petitioned the nation's highest constitutional court, the Constitutional Tribunal, to declare the proposed Water Code reforms unconstitutional.[40]

The conservative deputies' position was extreme: they argued that it would be unconstitutional not only to impose fees for nonuse on existing water rights, but also to place additional limitations or conditions on new water rights that might be granted in the future. Either reform, in their view, required approval by the large legislative majorities that were required for matters of constitutional rank.[41]

If their constitutional challenge had been successful, it would have gutted even the weakest version of reform. The Constitutional Tribunal, however, rejected the challenge two months later and accepted several of the government's arguments about the constitutionality of the proposals in question.[42] The ruling is the tribunal's only decision to date in the area of water law. As DGA's chief lawyer later argued, the decision affirmed the legal status of water resources as public property ("national goods for public use") and concluded therefore that private access to *new* rights to use those resources could be subject to regulations. The decision also concluded that those regulations can be imposed by ordinary legislation—that is, by simple majorities in both houses of Congress. In other words, the tribunal ruled that private use rights that have not yet been granted do not enjoy constitutional protection from new regulations.[43]

Although this was undoubtedly a victory for the government, the scope of the tribunal's decision was in fact quite narrow. It allowed ordinary legislation to impose limitations on future rights, but it did not clarify what limitations—including fees for nonuse—could be retroactively applied to existing rights. In practical terms, the latter question is obviously far more important.

The Senate commissioned a report on that latter question by José Luis Cea, a prominent constitutional law scholar at the Catholic University. Cea concluded that the government's proposals and arguments were indeed constitutional.[44] However, while Cea's opinion carries some influence in Chilean legal circles, his interpretation is of course not binding on the courts or on the Constitutional Tribunal. The Senate also commissioned a more technical legal analysis of the government's proposals by the Institute of Mining and Water Law, which is affiliated with the University of Atacama. Although the author of the institute's report agreed with the need to reform the Water Code, he doubted the constitutionality of some of the proposals' main features, including the imposition of fees on existing water rights.[45]

Another and equally prominent constitutional law scholar at the Catholic University, Raúl Bertelsen, has reached the opposite conclusion from Cea. Bertelsen, who is more conservative than Cea, argues that the proposed fees for nonuse are unconstitutional for several reasons. First, the fees "weaken the essential attributes or faculties of ownership held by the owners of water rights" and "affect the essence and free exercise of [their] property." Second, the fees "establish manifestly disproportionate or unjust taxes." Third, with respect to the proposed differentiation between consumptive and nonconsumptive rights, the fees "weaken the constitutional guarantees of equality before the law and, especially, [represent] arbitrary discrimination on the part of government in economic matters."[46] Since Bertelsen was a member of the military government's Constitutional Commission in the 1970s and participated in the drafting of the 1980 Constitution, his interpretation of constitutional intent must be taken seriously.

DÉJÀ VU ALL OVER AGAIN? THE REVIVAL OF WATER RIGHTS TAXES

Although the Constitutional Tribunal's 1997 decision did not go as the antireformers had hoped, the political debate dragged on for the next six years with little change on either side. The proposed reforms moved slowly through the Senate, where they were studied by three committees as well as debated on the Senate floor in late 2000.[47] The government made minor concessions and adjustments along the way but otherwise managed to keep the bill alive and moving forward.

The annual Chilean Water Law Conference has illustrated how slow and difficult the legislative process has been. This conference has been held at the Catholic University Law School in Santiago every November since 1998, providing a national forum for public and academic discussion of water law and policy issues. The need for such a forum had become evident by the late 1990s.[48] Despite its name, the Water Law Conference has brought together water specialists from many disciplines and organizations, including high-level managers from DGA, other Chilean government agencies, and private sector interest groups, as well as academics. Every year since 1998, the conference has included papers and panel discussions about the state of the Water Code reforms, generally focusing on the issue of fees for nonuse, and every year the presentations and arguments have been remarkably similar on all sides. To the outside observer, the discussions have often seemed like ceremonial repetitions or public rituals rather than policy debates, and regular attendees are tempted to check the calendar to remind themselves what year it is.

Since 1996 the alternative suggested by the reform's opponents—besides their preferred option, no reform at all—has been the now familiar notion of water rights taxes. These taxes would be like property taxes on land or real estate: the owners of water rights would pay the tax every year, whether or not the rights were used. Water rights taxes would be separate from land taxes, and taxes on irrigated land would be reduced accordingly so that the total tax burden on irrigated land would not change. This system is the same as the system that had been proposed by the military government's neoliberal economists in the late 1970s (as discussed in Chapter 2).

The arguments made in favor of water rights taxes in the late 1990s echo those made 20 years earlier. The first is that a system of taxes would strengthen the legal security of private water rights by reinforcing their status as real property, at the same time reinforcing the owners' freedom to use their property as they choose. In contrast, a system of fees for nonuse would increase government scrutiny of private water use and would strengthen the idea that water resources are public property to which private rights holders are granted access. The second argument is that taxes would provide more direct economic incentives for water rights owners to either use their rights efficiently or sell the rights not being used, thereby clarifying price signals and encouraging market forces. As we saw in Chapter 2, this argument was originally said to be the key to an effective water market, according to the military government's economists.

Those arguments serve to defend the current legal and economic framework, of both the Water Code and the Constitution in general, rather than call them into question. It is not surprising, therefore, that the strongest proponents of water rights taxes have been the firmest believers in neoliberal economic and ideological positions—in the late 1990s just as in the late 1970s. These proponents include the Institute for Freedom and Development, a neoliberal center of policy analysis that is closely tied to Chile's most right-wing political party, Independent Democratic Union, and *El Mercurio*, Chile's most influential and conservative newspaper.[49]

What appears more surprising, in light of Chilean farmers' hostility to water rights taxes, is that two influential spokesmen for agricultural water users have been among the leading proponents of such taxes. They are Luis Simón Figueroa, the lawyer whose warning about government planning on the attack was quoted above, and Fernando Peralta, an engineer who since the 1980s has been the president of the Confederation of Chilean Irrigators (the country's largest trade association and interest group for irrigators). Part of their argument is that under the present system, farmers are the only water users who already pay taxes on water rights, albeit indirectly through land taxes, because irrigated land is much more valuable than nonirrigated land and is appraised accordingly.[50]

Despite their positions as agricultural sector spokesmen, however, few farmers support Figueroa's and Peralta's argument. The explanation for this apparent paradox is that both men are firm believers in the Water Code's free-market principles, and defending the free-market model is a higher priority than protecting the immediate or perceived interests of farmers. Furthermore, both men are politically astute and are well aware of the rhetorical power of water rights taxes, as discussed below.

In addition to the more dogmatic neoliberals, other Chilean economists have also recently raised questions about the economic logic and efficiency of the proposed fees for nonuse. In 2000, both in the Senate Finance Committee and later on the Senate floor, senators discussed the perverse incentives created by charging a fee for *not* using a resource: surely, people argued, this would encourage economic inefficiency by encouraging wasteful water use.[51] Several participants in this debate had been trained as economists, and their concerns were shared by members of the Concertación as well as by the opposition.

The discussion in the Senate was influenced by a paper written by two respected economists at the University of Chile, who analyzed the government's proposed reforms at the request of a progovernment senator. These economists argued that the proposal was "suboptimal" and that a "definitely more convenient" solution would be to levy taxes on water rights whether or not they were used.[52] They also criticized the government's position that the concentrated ownership of nonconsumptive water rights in the hands of certain electric utility companies was a reason for reforming the Water Code. Instead, they argued, such problems should be addressed through Chile's antimonopoly laws. These criticisms were an especially hard blow to the government's case because the two economists in question were considered politically sympathetic to the Concertación.[53]

In response, DGA could only repeat the arguments that have been made so many times over the previous 20 years. The agency said that a system of water rights taxes was attractive in theory but too hard and too costly to implement in practice, at least in the short term. Such a system would require determining the value of hundreds of thousands of water rights throughout the country; creating and maintaining a database of all the legal titles to those rights;[54] reappraising the value of irrigated land throughout the country; and overcoming the political resistance of many thousands of water users, both irrigators and others. These were the same administrative, technical, political, and budgetary obstacles that had daunted the military government 20 years earlier, as discussed in Chapter 2. In contrast, fees for nonuse would apply to a much smaller universe of water users and therefore would be more feasible to implement—particularly because the government's real priority was nonconsumptive water rights, and it was willing to design the reform to effectively exempt the vast majority of consumptive rights holders.[55]

PROS AND CONS OF THE ECONOMIC INSTRUMENTS

In sum, there have been plausible arguments both for and against each of the proposed economic instruments: charging fees for the nonuse of water rights versus levying taxes on water rights ownership whether or not the rights are used. Both alternatives would involve redefining the rules affecting property rights to water in order to improve the economic incentives for water use and allocation.

To recapitulate, the government's arguments in favor of fees for nonuse instead of water rights taxes have been the following:

- Fees for nonuse are a first step in reform that could be implemented in the short term, would improve the efficiency of water use, and do not rule out more ambitious and sophisticated economic instruments in the future.
- Taxes or other charges for water use would be more difficult and complex to implement and would require stronger political will, more technical and administrative capacity, and a larger budget.
- A fairly broad social and political consensus exists that speculation and hoarding are unacceptable features of the Water Code, and fees for nonuse address those features directly.
- Fees for nonuse would impose fewer restrictions on existing water rights and thus raise narrower constitutional issues.
- In practice, fees for nonuse would affect nonconsumptive water rights much more than consumptive rights, and therefore agricultural interests should not object and implementation should be relatively manageable.

The arguments for water rights taxes, made by the government's political opponents and by other economists, have been the following:

- Fees for nonuse would not in fact be easier to implement than taxes because both measures require similar technical and legal information and administrative capacity.
- Fees for nonuse would create perverse incentives that encourage inefficient use of water rights, whereas taxes would provide clear economic incentives for more efficient water use and allocation.

- Taxes would strengthen the legal security of water rights by reinforcing their status as private property.
- Taxes would benefit farmers, currently the only water users who already pay taxes on water rights, indirectly through land taxes.
- If nonconsumptive water rights and monopoly power in the electricity sector are the primary target of the reform, then the reform should be designed to reach that target directly rather than tinker with the overall water rights system.

THE BOTTOM LINE

The bottom line, however, is this: even if we accept the criticisms of fees for nonuse as well founded, as I do myself, water rights taxes in Chile are a false alternative—appealing in theory and rhetoric but impossible in practice, at least for the foreseeable future. The same factors that scuttled these taxes in the late 1970s still exist. To implement a system of water rights taxes would demand a massive and nationwide political, administrative, legal, and technical effort, first to establish the system and then to maintain it. Creating and operating such a system would require as much administrative discretion as a system of fees for nonuse, despite some antireformers' arguments to the contrary, and hence would be just as open to bureaucratic abuse or corruption.

Although neoliberals and some other economists favor these taxes for theoretical reasons, there is no reason to believe that right-wing politicians or private sector interest groups would in fact put their weight behind such an ambitious and controversial reform; indeed, it would be little short of miraculous. It is hard to imagine what compelling political pressure could make current property owners act against their own immediate material interests. It is just as true now as it was 20 years ago that Chilean farmers would reach for their shotguns rather than pay new water rights taxes, notwithstanding the argument that such taxes would benefit agriculture in relation to other water-using economic sectors. For big nonagricultural water users, such as mining companies and electric utility companies, the status quo has been highly beneficial, even if they sometimes admit that it is based on bad economic theory. Moreover, conservative politicians who have already challenged the constitutionality of imposing fees on future water rights would almost certainly challenge new taxes on existing rights, and with a stronger constitutional argument. In the face of all this opposition, today's democratic government has much less power to impose its will than the military government had 20 years ago.

In short, the notion of water rights taxes is a rhetorical device and political tactic rather than a real counterproposal. The obstacles to implementing such taxes are so high that it is hard to escape the conclusion that at least some of their advocates are being deliberately insincere, if not hypocritical. As DGA's head lawyer pointed out in late 2001, to insist on taxes instead of fees for nonuse, in present-day Chile, is really to oppose any reform at all.[56]

Reforming Chile's Water Code after the country's return to democracy has proven much more difficult than the Chilean government had expected and than foreign observers have supposed. After more than 10 years of sustained effort, both the terms of debate and the scope of the proposed reforms have narrowed considerably. There has been general political agreement about the need to change the economic incentives for water use, which would require some change in the legal rules that define water rights, but so far it has been impossible to reach agreement on the specifics. The fate of even the most recent and modest version of the proposed reform remains uncertain.

I return to these issues in the concluding chapter, where I discuss their significance and implications for the overall evaluation of the Chilean water model. Before that, however, in Chapter 4 I look at the evolving understanding and research about the model's empirical results.

CHAPTER 4

THE RESULTS OF CHILEAN WATER MARKETS: EMPIRICAL RESEARCH SINCE 1990

The market for water rights has been the feature of Chile's 1981 Water Code that has attracted by far the most attention. This is especially true in international circles, where the Chilean Water Code has been perceived as essentially synonymous with Chilean water markets and water rights trading. Other issues of water management and the current institutional framework have been distinctly secondary in academic and policy research. Moreover, most of the research to date has been done by economists, which helps explain the focus on markets and trading.

The performance and results of Chilean water markets have been highly politicized topics both inside and outside Chile. This is not surprising: the Water Code is so pure a symbol of free-market theory and ideology that both proponents and critics have had a lot at stake in whether the resulting markets are considered a success. Empirical and academic research about these markets has lagged far behind the policy debates and has evidently been a much lower priority for most people involved.

Within Chile the issue has been additionally politicized because it is so closely related to the national debate about reforming the Water Code (see Chapter 3). Both sides in that debate have had strong motives to support their positions with claims about how well or how poorly the water market has worked in practice. In the international water policy arena, the topic has been politicized since the World Bank began to publicize the Chilean Water Code as a model of successful reform in the early 1990s (see Chapter 1). In Latin America, the Inter-American Development Bank has tended to follow the World Bank's lead in this area, since it has less research and analytical capacity than its larger sister organization. This provoked a strong critical reaction from water experts at other international organizations, particularly United Nations agencies. The UN Economic Commission for Latin America and the Caribbean, which is in Santiago, was especially active.

Because of the lag in empirical research, much of the discussion about Chilean water markets has been long on theoretical or ideological argument and short on reliable information. This was especially true in the first half of the 1990s. The amount of research increased slowly throughout the 1990s, however, and by the end of the decade the level of empirical understanding had improved significantly. It is notable that most of this empirical research has been done by foreigners rather than Chileans. Although this is partly because funding for research is limited in Chile, a more important factor has been the politicized nature of the national policy debate. With few exceptions Chilean analyses have been driven by the government's proposed legislative reforms—whether in favor or against—rather than by the goal of creating original and empirical knowledge for public discussion.[1]

In this chapter I review the evolution of empirical research since 1990—that is, how water markets in Chile have worked in practice rather than in theory.[2] Published accounts have changed in tone and content from the early 1990s to the late 1990s, from exaggerated claims of success to more balanced assessments of mixed results. By the second half of the decade a good deal of consensus had emerged about the main characteristics of Chilean water markets, at least among people who were well-informed, although disagreement remained about the overall assessment and the policy implications that should follow.

The growing consensus about the empirical results and basic facts of Chilean water markets points to a different problem: researchers have so emphasized the buying and selling of water rights that they have virtually ignored other issues that are at least as critical to integrated water resources management. These missing issues include the impacts on social equity, coordination of multiple water uses, river basin management, resolution of water conflicts, and environmental protection. Moreover, nearly all the research on Chilean water markets has focused on *consumptive* water rights. Although nonconsumptive water rights have presented serious problems of private speculation, monopoly power, and river basin conflicts between farmers and hydroelectric companies, these rights have rarely been traded in markets. In consequence, nonconsumptive rights have been left out of most empirical studies.

In the latter part of this chapter, I discuss the missing issues and the available evidence about how they have fared under the Chilean water law model. I will argue that the obsessive focus on water rights trading, to the exclusion of other issues of water management and institutions, reflects the predominance of narrow economic perspectives among those who have examined the Chilean experience.

The works reviewed in this chapter include essentially all the relevant research published in English and Spanish. My purpose here is to highlight the general trends and features of this research, not to provide an exhaustive summary. For more details readers should consult the works cited.

THE OVERALL TREND: FROM PARTISAN BOOSTERS TO GREATER BALANCE

The general trend over the course of the 1990s was from exaggerated claims about the remarkable success of Chilean water markets toward more moderate and credible assessments of their mixed results. During the first half of the decade, published accounts tended to be highly enthusiastic. These accounts' ideological bias and lack of empirical foundation were evident from their sweeping assertions of success on all counts: the water market had supposedly resulted in active trading of water rights, greater efficiency of water use and allocation, social and economic benefits for poor farmers, and fewer water conflicts. No significant problems or difficulties were recognized.

All of these early publications were by economists associated with the World Bank, either as staff members or consultants. It is important not to oversimplify the World Bank's position on water policy and economics.

Although the Bank is a large organization, it is not monolithic, and it includes people with a certain range of viewpoints. With respect to Chilean water markets, the quality of the Bank's analyses and publications varies a good deal: some are reasonable and well supported, others are misleading or simply wrong. Unfortunately, it is often hard for a reader to tell the difference, particularly because most of these publications share a tone of confident assertion. In any event, despite the variations, these publications share a narrow economic perspective, and their overall spin on the Chilean model is always positive.

Several examples of the early wave of exaggerated claims were published in 1994. One was a World Bank report about water management and irrigation development in Peru, written by Bank economist Mateen Thobani. Thobani played an active and visible role in publicizing the supposed benefits of the Chilean model in Peru and elsewhere. In this report he discusses a new draft water law for Peru that had been modeled on the Chilean Water Code; indeed, the draft was written by Chilean consultants who had been part of the team that wrote the Chilean law in the late 1970s.

Thobani's description of the Chilean system is selective and misleading. According to Thobani, the Chilean water law "has successfully improved water delivery and use, stimulated private investment, and reduced water conflicts." In addition, he asserts that the Chilean law has increased the value of water, reduced environmental damage, and benefited poor farmers at the expense of "politically influential water users." He does not provide or cite any evidence to support these assertions, which appear to be based on theoretical arguments about property rights.

At the same time, Thobani's report argues in favor of several crucial provisions in the draft Peruvian law that were in fact very different from the Chilean model, although he does not mention the differences or explain their significance. For example, the Peruvian draft law was based on a system of property taxes on water rights, public auctions of new water rights, and strong regulation of monopoly power over nonconsumptive rights, and it included a public information campaign to discuss the new law. All of these provisions were missing in Chile and had already been shown to be politically or administratively infeasible there, but they were presented to Peru as if they were simply part of the Chilean package.[3]

Another example is a 1994 paper written by Mark Rosegrant, an economist at the International Food Policy Research Institute in Washington, and Hans Binswanger, an economist at the World Bank. In this paper (already discussed in Chapter 1), Rosegrant and Binswanger present a comprehensive argument in favor of markets in tradable water rights in developing countries, and they refer repeatedly to Chile to support their case. In Chile, they say, such water rights markets "have been operating effectively with relatively unsophisticated conveyance technology" and have "greatly reduced the number of water conflicts reaching courts."

Rosegrant and Binswanger address potential problems of social inequity and use the example of Chile to dismiss the concerns. They mention that some people have worried that differences in wealth or power might favor large nonagricultural water users and harm small farmers, if water rights are made fully tradable. They conclude, however, that "evidence from Chile, where active markets exist ... shows that this has not happened." They also refer to concerns raised about the influence of market power in the initial assignment of water rights; here too, they say this was not an issue in Chile because water rights had been assigned as part of the military government's reversal of land reform, which "was seen as an improvement in equity."[4] (This last claim is simply mistaken: the great majority of water rights in Chile were *not* in fact assigned as part of the end of the Agrarian Reform, as I discuss in Chapter 2 and again later in this chapter. This error, however, has been repeated routinely in later publications and has become one of the common myths about the Chilean case. Similarly, readers should not accept at face value the assertion that the military's redistribution of land and water rights in the 1970s made small farmers better off, or that it was perceived to have done so in Chile. Such a statement is debatable at best, and in Chile it would be understood as a political opinion rather than an objective comment.[5])

In short, Rosegrant and Binswanger's description of Chilean water markets is uniformly rosy. This reflects their principal source of information: they rely heavily on the work of Renato Gazmuri, a Chilean economist and politician. Gazmuri is knowledgeable and experienced, but he is hardly an impartial observer. He was a high-level civilian official in the Ministry of Agriculture in the military government during the mid to late 1970s, when he was a member of the neoliberal team that reversed the Agrarian Reform, liberalized the agricultural sector, and designed the 1981 Water Code. As discussed in Chapter 2, that was a highly ideological period and process. During the early 1990s Gazmuri worked as an international water policy consultant in several other countries (including Mexico, which passed a new water law in 1992 after considering and rejecting the Chilean model). Gazmuri joined Rosegrant to publish several papers that spread the word about the Chilean "success story."[6]

The viewpoints summarized above were the first to be published in English, and in international circles they dominated the initial terms of debate. They soon began to unravel, however, as more careful empirical studies were completed. From 1995 on, growing evidence from both Chilean and foreign researchers—including some who were funded by the World Bank—led to more balanced assessments of the limitations of Chilean water markets.

One area where the conventional wisdom shifted was the question of whether Chilean water markets were "active," as their early boosters contended. The first empirical study to challenge that assertion was a paper I published in Chile in late 1993, which discussed the results of two years of field research in Chile for a Ph.D. in law and political economy at the University of California–Berkeley. That this paper was published in Spanish limited its circulation outside Latin America, although in Chile it had a lasting impact. This research became available in English in 1995 and was later published in expanded and updated form in 1997–1998.[7]

In these publications I argued that the available evidence, both quantitative and qualitative, showed that water rights transactions were in fact quite uncommon in most parts of Chile, and therefore, as a general rule, Chilean water markets were relatively inactive. Moreover, the great majority of water rights transactions took place *within* the agricultural sector and did not involve nonagricultural water uses. These were empirical observations rather than a criticism of the water market, and much of my analysis sought to explain the markets' observed inactivity by discussing the many factors that limited water rights transactions.

These limiting factors include (in no particular order of importance):

- constraints imposed by physical geography (Chilean rivers are short and steep and interbasin transfers are expensive) and by rigid or inadequate infrastructure (e.g., canals with fixed flow dividers and very few storage reservoirs);
- legal and administrative complications, particularly the uncertainty and confusion about water rights titles and recordkeeping;
- cultural and psychological resistance to treating water as a commodity, especially from farmers; and
- inconsistent and variable price signals about the real scarcity and economic value of water (e.g., water rights owners are rarely willing to sell, even if their rights are unused, and until recently groundwater has been an untapped alternative).[8]

The first two factors in particular—problems of infrastructure and legal title—have been common themes in all subsequent analyses of Chilean water markets, as discussed below.

The activity of water markets, in the number or frequency of water rights sales, was only one way to assess the Water Code's results, and not necessarily the most important. A more critical issue was the effectiveness of the law's market incentives in increasing the efficiency of water use and allocation, specifically by encouraging investment in water conservation so that the water saved could be sold. This was the main economic argument in favor of the new Water Code back in the late 1970s, as described in Chapter 2.

I argued that these market incentives had been almost entirely ineffective in practice. Water rights owners in Chile rarely sell any unused or supposedly "surplus" rights; instead, they hold onto such rights to protect themselves from occasional drought years or because they know that the value of those rights will increase over time. Even where farmers have invested in more efficient water use, their motive has been to improve their agricultural yield or to expand their irrigated acreage, and they have not sold any resulting surplus water. Indeed, since 1985 the Chilean government has reverted to subsidizing private investment in irrigation in an implicit recognition that, contrary to the intent of the economists who wrote the Water Code, the water market has not provided sufficient incentives. The ineffectiveness of these incentives, of course, was partly due to the absence of any taxes on water rights or other costs of ownership, as the neoliberals had argued to no avail in the late 1970s.[9]

My arguments in this research were considered "antimarket," particularly in Chile, but this reflected the politicized nature of the debate rather than the research itself. In fact, my criticisms were aimed not at Chilean water markets but rather at the exaggerated claims being made about their success. I agreed that the markets had the advantage of allowing flexible reallocation of water resources, even if this advantage was still more potential than real in the mid 1990s. I predicted that water markets would become more active over time and in certain regions of Chile, as water demands and relative water scarcity increased enough to overcome the obstacles and transaction costs listed above. In short, I concluded that the most important economic benefits of the Water Code have come not from water rights trading or market incentives, but instead from the greater legal security of property rights, which has encouraged private investment in water use.

The next empirical research was done for another Ph.D. dissertation, this time in agricultural economics at the University of Minnesota. Robert Hearne studied four areas in central and northern Chile, selected precisely because they were expected to have active water markets. In each case the climate was arid, water was scarce, and irrigated agriculture was well developed and commercially profitable (the southernmost of the four cases included part of the Santiago metropolitan area). Hearne's research showed, however, that there was very little trading of water rights in three of the four study areas. The principal explanation was that the rigid canal infrastructure made it costly to change water distribution, particularly among farmers. The one exception was the Limarí River basin in north-central Chile, which I discuss further below since it has become widely known as *the* example of Chilean water markets (see Map 1 on page iv).

Hearne argued that the water market had led to economic benefits in some areas, including greater economic efficiency due to transfers to higher-value water uses—that is, from agriculture to urban supplies. (Even here, however, the transfers involved mainly paper titles to water rights that had been long unused, rather than a physical reallocation of water resources.) In general he favored the Chilean model. Nonetheless, his empirical work helped confirm the view that water rights sales and transactions were the exception rather than the rule in most of Chile. His conclusion carried added weight because the World Bank had financed his research and he himself was a World Bank consultant at the time.[10]

That conclusion soon became part of the conventional wisdom about Chilean water markets. It forced the Water Code's proponents, both Chilean and foreign, to change their arguments about the code's success. From 1995 on, many of the advocates no longer said that the water market was active; instead, they argued that even though the market was inactive, it was nonetheless efficient. In fact, they said, the lack of water rights transactions showed that the existing allocation of water resources was already efficient. This became a common argument in Chilean debates about reforming the Water Code (see Chapter 3).

Chilean consultants Mónica Ríos and Jorge Quiroz repeated this argument in another World Bank publication in 1995, in which they reviewed the major issues raised by Chile's market in water rights. Ríos and Quiroz did not conduct any new empirical research: their review is based on interviews and on the limited existing literature summarized thus far in this chapter. Their review is fairly balanced, and their analysis is similar to Hearne's. Ríos and Quiroz question the significance of the lack of water rights sales, for the reason mentioned just above, and argue instead that the water market has been active in temporary rentals. This assertion is probably accurate, although they offer no evidence for it, since farmers in the Chilean countryside have long made temporary exchanges of water rights. Such exchanges are typically informal arrangements and are facilitated by local canal associations, which deliver the water. In any case, there is no reason to attribute these rentals to the current Water Code, at least until empirical research shows otherwise.

Overall, Ríos and Quiroz conclude that "the system in Chile has worked reasonably well" and that the problems identified should be addressed through "'fine tuning' of the system rather than drastic reform." The problems they describe include the vague definition of nonconsumptive water rights, which caused conflicts with consumptive rights, and the "transaction costs arising from incomplete legalization of water titles, lack of infrastructure, and free rider problems." Among the "minor amendments" they recommend is a user fee on all water rights (in 1995 the political difficulty of such a "minor" change had not yet been demonstrated, as discussed in Chapter 3).[11]

The Chilean government responded to the change in conventional wisdom by making the inactivity of the water market one of the major reasons for reforming the Water Code. In its 1996 package of proposed legislative reforms, the government argued that the Water Code's laissez faire definition of water rights—that is, including neither financial costs nor legal obligations—had distorted price signals and economic incentives and hence had distorted the market. The reforms would allegedly make the market more dynamic and more efficient (see Chapter 3). The Dirección General de Aguas (DGA) developed this analysis of the market's flaws in its National Water Policy and related agency documents from 1996 on, as discussed below. DGA conducted internal studies of water rights transactions to confirm these arguments about market performance but did not publish additional empirical evidence for years.[12]

In international circles, the conventional wisdom about Chile was also affected by the counter-offensive led by the UN Economic Commission for Latin America and the Caribbean (ECLAC) in the mid 1990s. Water experts at ECLAC and other international organizations were deeply concerned about the way the World Bank and Inter-American Development Bank were pushing the Chilean model on other developing countries, both in Latin America and elsewhere. These skeptics included some staff people at both Banks. ECLAC staffers Miguel Solanes, an Argentine water lawyer, and Axel

Dourojeanni, a Peruvian engineer and specialist in river basin management, were especially active in mobilizing opposition. Although they conducted little new empirical research themselves, they assembled the available information to highlight the flaws of the Chilean model, and at international meetings and workshops as well as in published articles, they urged other countries to take a more balanced approach.

ECLAC officials and other international water experts were familiar with economic arguments and favored the appropriate use of market incentives, but they did not share the more dogmatic neoliberal views of many proponents of the Chilean model. For example, Solanes argued that Chile was the only country in the world that did not impose any legal conditions on private water rights, and that this ran counter to hundreds, if not thousands, of years of international experience. In his view the essential point was that water rights had to be subject to some requirement of socially beneficial use, which could be enforced by government authority; otherwise, the danger was too great that public interests in water management would be harmed by monopoly power, unfair competition, speculation, hoarding, and environmental damage. Solanes argued that the lack of this beneficial use requirement in Chile had led to precisely those problems, particularly with respect to nonconsumptive rights and the electric power sector.[13]

After 1995, no new empirical research became available until the end of the decade, although some new publications appeared that were based on existing research.[14] In the meantime, with the description of the market's inactivity widely accepted, researchers in Chile began to develop more sophisticated diagnoses of the reasons for that inactivity and the range of problems needing attention. Nevertheless, during this period of the later 1990s, various economists associated with the World Bank continued to publicize the benefits of the Chilean model, as discussed in Chapter 1.[15]

RECENT CHILEAN OVERVIEWS: TOWARD A SHARED DIAGNOSIS

In this section I summarize several important syntheses and overviews about Chilean water markets that have been published in Chile since 1997. These publications are available only in Spanish and are largely unknown outside Chile. They represent the state of knowledge of the best-informed Chilean experts. In most respects these papers also confirm the analyses presented in my own research a few years earlier, with which these Chilean scholars were familiar.[16]

In 1997–1998, Chile's leading academic expert on water law, Alejandro Vergara, published several articles that examined the functioning of the water market. Vergara is a law professor at the Catholic University in Santiago as well as a water lawyer in private practice. His analysis was based on a thorough review of the existing literature and on his own professional experience. Vergara's point of departure is that Chile has already adopted water rights legislation that favors the free market rather than government planning, and he does not question or criticize that decision. Instead his purpose is to raise issues about how the water market has actually worked so far and suggest the legal improvements needed for it to work better in the future.[17]

Vergara's initial observation is that "the free market was established but not all of the prior institutional arrangements were made that are necessary for the market to function adequately."[18] The gist of his argument is that water rights in Chile are *not* in fact clearly defined, despite the Water Code's general principle in favor of private and tradable property rights. In both legal and physical terms, water rights are much more fuzzy than they seem on the surface. He identifies the following problems:

- The Water Code does not deal with the externalities caused by water rights transactions, either environmental impacts or third-party effects on other water users. Indeed, both Vergara and other defenders of the Water Code have argued that externalities have not yet been a problem in Chile precisely because the water market has been so inactive.[19]
- A great deal of legal confusion surrounds the rights to return flows from upstream irrigators. According to the Water Code, downstream users have no rights to those flows, despite many decades of customary practice to the contrary.
- The recordkeeping and registration of water rights titles are widely recognized to be completely inadequate. The great majority of water rights in Chile are not formally registered; instead they predate the 1981 Water Code and are based on past legislation or on customary practice. Although these rights often lack any documentary support, they enjoy full constitutional protection as property rights. Unregistered rights cannot be bought or sold, however—another obstacle to water markets.[20]
- Many water rights that do have formal legal titles have a substantive content that is disputed in practice. For example, they may be defined as "permanent" rights when in fact the water is not available all year.

- The existing infrastructure of canals and reservoirs is inadequate to allow many transfers of water from one place to another.

Vergara argues that until those problems are addressed, water markets will remain restricted and flawed and their potential will not be realized.

Guillermo Donoso is a colleague of Vergara's at the Catholic University and the leading Chilean academic economist in the field of water markets. He is a strong supporter of water markets, and from the standpoint of neoclassical economics he has assessed how these markets have worked in Chile and what factors have prevented them from working better. In the mid 1990s Donoso's analysis was almost entirely theoretical, as if drawn from an economics textbook, but by the end of the decade he had also done more empirical work.[21]

At the first Chilean Water Law Conference, in 1998, Donoso presented an overview of the functioning of the water rights market and an identification of its problems, based on a review of the academic literature and various consultants' reports.[22] His account echoes Vergara's and mine in most respects. He begins by summarizing the conflicting views about how active Chilean water markets have been, concluding that although there is some room for debate, it is clear that the markets exist to some degree but vary greatly by river basin and geographic region: the market is more active in areas where water is scarcer and during periods of drought.

The bulk of Donoso's paper considers the various problems and distortions that have affected Chilean water markets. He first discusses problems that would affect any system of water rights allocation, whether a market or administered by government. Like Vergara, he argues that the major problem is the inadequate definition of water rights, which has caused negative externalities when rights have been sold or transferred, including loss of or interference with return flows, degradation of water quality (though to some extent this problem falls outside the jurisdiction of the Water Code), and overextraction of groundwater. He also describes the unavoidable transaction costs caused by the need to build or modify physical infrastructure to redistribute water.

Donoso then describes the problems that are particular to the market. The worst such problem, in his view, is the lack of adequate legal, economic, and hydrological information about water rights. Although this is also a problem for nonmarket systems of allocation, he argues that it is more acute for a market because a decentralized system relies more heavily on good and widely available information. Closely related problems are the gap between nominal rights (i.e., paper rights) and real rights (sometimes called "wet water"), and the conflicts caused by transactions involving or affecting the many thousands of unregistered customary rights. Finally, he discusses speculation and hoarding of both consumptive and nonconsumptive rights, which he concludes are minor issues.

In a later paper, Donoso presents quantitative data about water rights transactions in two river basins, the Maipo and the Limarí (both studied by Hearne in 1993). Since the Maipo River basin includes metropolitan Santiago as well as irrigated agriculture, many of the transactions involve urban water supply companies or real estate developers buying water rights from farmers (see Map 1). The data show that only a very small percentage of existing water rights are sold each year and that prices vary widely, but some reallocation has taken place. The market in the Limarí River basin is more active and functions better, for reasons discussed below. Although Donoso's paper does not contain arguments or results that challenge what is already generally known in Chile, it is nonetheless notable simply for its basic empirical description and analysis.[23]

Finally, in 1999 two water experts at ECLAC published a comprehensive analysis of the Chilean Water Code, suggestively subtitled "Between Ideology and Reality."[24] Axel Dourojeanni and Andrei Jouravlev argue that too many other countries in Latin America have looked to Chile as the model of water law and policy reform without knowing the problems that the Water Code has caused or the controversies within Chile about how to solve them. Dourojeanni and Jouravlev attempt to rectify this omission in a long and substantive paper that is abundantly sprinkled with quotations from the many works cited. Their paper is aimed primarily at Latin American readers outside Chile.

First, they describe the problems with the original allocation of water rights: speculation, accumulation and hoarding, and excessive monopoly power. They argue that these problems are serious in the case of nonconsumptive rights and the electricity sector, and relatively unimportant in the case of consumptive rights and agriculture. This chapter also describes the government's proposed Water Code reforms aimed at correcting these problems.

Second, Dourojeanni and Jouravlev analyze why water markets have been so inactive in Chile and why water rights transactions have been so uncommon. Here they repeat the analyses of prior studies, discussed above.

Third, they examine the problems caused by the Water Code's inadequate regulation of externalities, both in the government's original granting of water rights and in the subsequent transfers of water rights. Their discussion includes several kinds of externalities:

- return flows—that is, the relations between upstream and downstream water uses and water rights;
- in-stream effects, including in-stream flow protection for environmental purposes as well as the coordination of extractive and in-stream water uses (i.e., consumptive and nonconsumptive uses); and
- the impacts on "areas of origin," meaning areas whose water supplies are sold or exported; here their concern is mainly for rural farming communities and indigenous communities.

In this context Dourojeanni and Jouravlev also describe the weaknesses of existing institutional procedures for reviewing third-party effects and resolving conflicts.

Fourth, the paper concludes with a brief summary of the Water Code's economic achievements. The authors' main purpose, however, is to counter the generally positive images that have dominated most descriptions of the Chilean experience, and in this they differ from Vergara and Donoso, who support the Water Code even as they recognize some of its problems.

THE LIMARÍ RIVER BASIN: POSTER CHILD FOR CHILEAN WATER MARKETS

One particular river basin in north-central Chile attracted ever more national and international attention through the 1990s: the Limarí River and its tributaries (see Map 1). The Limarí River basin is the one example that is widely agreed to have an active and successful agricultural water market, including both temporary rentals and permanent sales, and even local real estate agents who broker and facilitate water rights trading. The Limarí basin is the place that foreign economists come to study and that the Water Code's boosters prefer to talk about.[25]

The Limarí basin has three factors in its favor that are not combined anywhere else in Chile. First and most important, the basin has adequate water storage capacity, thanks to three reservoirs built by the national government between the 1930s and 1970s and maintained and administered by the Ministry of Public Works. These reservoirs are for irrigation purposes only. Second, the local water users' associations are for the most part well organized and their canal infrastructure is well maintained. Third, the climate is sunny, hot, and dry, making excellent conditions for growing high-value fruit crops for export. For all these reasons, the water rights market is dynamic. It is important to keep in mind, however, that irrigation is by far the most important water use and that the agricultural sector overwhelmingly dominates water rights trading.

The Limarí basin was the subject of a third U.S. Ph.D. dissertation about Chilean water markets, this time in agricultural economics at the University of California–Davis. This study, completed in 1999, is the most careful and sophisticated empirical research done on this topic since the first half of the 1990s. Ereney Hadjigeorgalis's study illustrates the intricacies of defining water rights in practice, as described above, and presents a rich and detailed description of a complex system of three reservoirs and their associated canals, known as the "Paloma system" after the name of the largest reservoir. She examines both short-term transactions (the spot water market) and permanent water rights sales. Her empirical analysis is obviously based on intensive fieldwork of a kind far too rare in Chilean water issues.[26]

In discussing the water rights in the Paloma system, Hadjigeorgalis makes a crucial distinction between their physical location and their legal source: for a given water right, different reservoirs may be involved. This is because the Paloma Reservoir—the largest as well as the last of the three reservoirs to be built—was built to increase the irrigation security of the valley by unifying and integrating the preexisting water distribution systems. The result was that a large number of water rights that formerly were supposed to be delivered via rivers or stored in the two smaller reservoirs were relocated to the Paloma Reservoir (in other words, the water corresponding to those rights was physically transferred). The distinction between the physical location and the legal sources of water rights leads Hadjigeorgalis to distinguish between physical and institutional constraints on water rights trading. Institutional constraints are defined as those imposed by the canal users' organizations to prevent third-party impacts.

Short-term trading is allowed between farmers within the same physical sector of the system—that is, between those who share the same reservoir, regardless of the legal sources of the water rights involved. In the permanent water rights market, by contrast, "the universe of possible trades is determined by both the physical *and* legal location of the water right." This is because physical constraints prevent transferring rights between reservoirs, while institutional constraints prevent trading rights "that are stored within the same reservoir, but that have different legal locations."[27] Hadjigeorgalis then presents a range of data about the nature of water rights transactions and price behavior, describing and comparing the spot markets and permanent markets in the different sectors of the overall reservoir system.

Hadjigeorgalis concludes that the Limarí water market has operated efficiently and has had important benefits for both buyers and sellers. There is abundant evidence that water has been frequently reallocated to higher-value uses within the reservoir system. In addition, the market has provided farmers with the flexibility to manage some of the risks caused by uncertainties in water supplies and in agricultural markets. Poor farmers, for example, have been able to lease their water rights to other farmers during drought years, when water prices are high and income from irrigation is uncertain. For these reasons, Hadjigeorgalis argues against the aspects of the proposed Water Code reforms that might penalize farmers for not using their water rights.[28]

As she herself points out, however, her study looks only at the water market within agriculture, and only at the country's most unusual case. In this sense the very success of the Limarí water market seems to be the exception that proves the rule.

ISSUES MISSING FROM THE RESEARCH

The work summarized above indicates fairly broad agreement about the empirical description of Chilean water markets, at least among people who are knowledgeable about them. This consensus is striking given that these authors have different theoretical and disciplinary perspectives and different positions on the Water Code in general. The consensus about empirical description does not extend to the policy implications, but the gradual accumulation of research in this area has sorted out the confusion caused by conflicting accounts about the basic facts in the first half of the 1990s.

What is equally striking, however, especially to the outside observer, is what has been missing from research about Chilean water markets—namely, the impacts on social equity, environmental sustainability, river basin management, coordination of multiple water uses, and resolution of water conflicts. These issues are often mentioned in political debates, albeit in general and rhetorical terms, and they are sometimes mentioned in passing in academic and policy research. But rarely have they been studied.

Most researchers, both Chilean and foreign, have focused their efforts on the economic and legal aspects of water rights trading. It bears repeating that the research I have reviewed thus far is essentially all of the relevant work in this area.

The absence of research on these social, environmental, and institutional issues is critical for two reasons: first, these issues are at the center of contemporary international debates about water policy reforms and integrated water resources management, as discussed in Chapter 1; and second, the available evidence suggests that these issues are in fact serious problems in Chile, as I argue in the rest of this chapter. That these issues have been secondary in research about the Chilean model points to one of the broader lessons to be learned from the Chilean experience: there has been too exclusive a focus on the economic aspects of water markets, and that focus has itself been too narrow.

CHILE'S NATIONAL WATER POLICY IN THE LATE 1990S

The Chilean government has been fully aware of the importance of the issues missing from research, as indicated by the more recent versions of DGA's National Water Policy. DGA began to prepare this policy back in 1990 as part of the initial Water Code reform package, as discussed in Chapter 3. DGA's work on this policy continued through the decade, closely related to the government's evolving reform proposals but also part of a broader process of modernizing the agency's organizational capacity and activities. The National Water Policy eventually took the form of a substantial document intended for public discussion and education in the late 1990s.[29]

This document is worth summarizing even though it has not been translated into legislative reform. The first half of *Política Nacional de Recursos Hídricos* presents background information, and the second half focuses on current policy issues and the proposed Water Code reforms. The document begins by stating five fundamental principles that are very similar to the Dublin Principles of integrated water resources management, described in Chapter 1:

- Water is legally defined as a "national good for public use," which reflects its vital importance to the public interest and therefore the need for some degree of government regulation.

- Water use must be environmentally sustainable.
- Water is an "economic good."[30]
- Water policy should promote participation by water users, other citizens, and social organizations, thereby deepening democracy and reflecting the social, economic, environmental, and cultural importance of water.
- The physical dynamics of water resources and the hydrologic cycle are extremely complicated, and water policy must recognize that complexity and be based on adequate scientific and technical knowledge.

After laying out those principles, the document translates them into a series of general policy objectives: to assure the supply of basic water needs, to increase the efficiency of water use and allocation, to reduce the impacts of hydrologic variability and environmental damage, and to minimize water conflicts. The document then describes the three major challenges facing water policy in Chile, illustrated with a variety of statistics and graphs. First is the steady increase in multiple demands for water in diverse economic sectors, generated by the nation's sustained economic growth since the late 1980s. Second is the growing pressure on environmental systems and worsening problems of contamination, again in diverse economic sectors. The third challenge is the uncertainty caused by climatic variability.

Following that general background, the document summarizes the essential features of the existing legal and economic framework: the private nature of water rights, the general application of free-market economics, and the concept of the "subsidiary state" (according to which government leaves as many activities as possible to the private sector, while retaining crucial regulatory tasks and promoting social equity).[31] There follows an outline of governmental agencies and functions that relate to the spheres of activity and responsibility of private owners and water users.

The second half of the document contains a more detailed diagnosis of current policy problems and the government's proposed solutions. This is where DGA explains and advocates the government's package of Water Code reforms. The issues fall into seven categories:

- the legal nature of water rights (with proposals such as establishing fees for nonuse, requiring justification for new rights, and reserving minimum ecological flows for new rights);
- institutional structure, integrated water management, and planning (especially the need to strengthen basinwide water management);

- environment and contamination (describing existing environmental damage throughout the country, fragmentation of government agencies, and the potential remedies offered by Chile's 1994 basic framework law for environmental protection);
- water resource use (including technical efficiency and also the operation of water rights markets);
- water administration and water users' organizations;
- scientific knowledge and information (including data collection and monitoring); and
- technical training and public education.

Some aspects of the National Water Policy have been questioned or criticized in Chile as part of the political debate about Water Code reform. Nonetheless, DGA's document is a substantial, thoughtful, and generally balanced contribution to public discussion. Whether or not it has much effect on legislation, it shows that the government's water experts have a good grasp of contemporary international water policy debates and the issues that need to be addressed in Chile.

MISSING RESEARCH ISSUE 1: SOCIAL EQUITY

The impacts of water markets on rural poverty and social equity in Chile have received very little research attention. There are several hundred thousand peasants and small farmers in Chile, generally referred to as *campesinos*, throughout the country. (A small proportion of *campesino* irrigators belong to indigenous communities in the country's northern desert and mountain regions, although the large majority of Chile's indigenous population lives in the wetter southern regions where land rights rather than water rights have historically been the major issue.) The water rights and water use problems that many of these farmers face have causes that predate the 1981 Water Code and reflect deeper problems of poverty and social inequality.[32]

The water rights problems of peasants and small farmers—both before and after 1981—can be summarized as follows. First, these farmers tend to lack secure water supplies or legal titles to water rights, although they may have customary or informal claims to use the overflow of more prosperous irrigators. Second, poor farmers' canal infrastructure and irrigation technology are more primitive and less well maintained than the national norm. Third, they have limited influence or voice within water users' organizations, which are dominated by larger farmers. Finally, peasants and small farmers tend to avoid the formal legal system, both courts and government bureaucracies, because it is generally costly, time consuming, and unsympathetic.[33]

The more empirical economic studies of Chilean water markets have looked at equity issues only peripherally, not as an important research question. According to both of the agricultural economics dissertations discussed earlier, for example, many small farmers in the Limarí River basin have sold their water rights to larger farmers or to agribusiness corporations,[34] but that result may or may not be inequitable: without further investigation we can only guess. Since the Limarí basin is Chile's best known, most dynamic, and most studied water market, one might have expected that someone would follow up these observations. So far no one has done so. Even Donoso, the most experienced Chilean economist writing in this area, leaves out social equity in his recent overview of the literature about Chilean water markets.[35]

Less empirically grounded economic analyses have tended to be more positive, contending either that the water market has made peasants and small farmers better off or, at worst, that any problems of social equity have been trivial. This is the view of most of the publications written by World Bank economists, discussed earlier in this chapter.[36] However, these authors have not shown any evidence in support of such claims, and their commentary in this area has lacked detail and depth.

For example, at the end of their overview of the "major issues" raised by Chilean water markets, Ríos and Quiroz include a brief paragraph about equity impacts. Referring to previous research that had raised the possibility of equity problems, they simply assert that "it seems to be a non issue in the case of Chile given the traditional operation of a water market among farmers." That is the extent of their discussion.[37]

Briscoe dismisses the issue in similar fashion, as part of his defense of the Chilean system as a model of how to manage water as an economic good. He quotes my work as an example of the argument that the Water Code has caused some problems for poor farmers, and he rebuts that argument by saying that the alternatives to water markets would be worse. Briscoe's rebuttal arguments are rather misleading, however. He cites proponents' contention that water markets have helped the poor by reducing government subsidies to irrigation, which had historically benefited more prosperous farmers, thereby reducing the national rate of inflation. These are dubious arguments, since in fact the military government continued to subsidize irrigation works after 1985—particularly for more prosperous farmers—and the Concertación governments have increased irrigation subsidies for both large and small farmers. In any event, it is unlikely that a small change in the inflation rate has been the most important effect of water markets on *campesinos*. Briscoe also argues that the Concertación has been "firmly committed" to water markets as well as to social equity and therefore has tried to overcome obstacles that have prevented the poor from taking advantage of them. Although this description of the Concertación's position is fairly accurate, if perhaps overstated, it confirms the idea that poor farmers have had problems with the current system.[38]

In contrast, most staff members in Chilean nongovernmental organizations (NGOs) and in the Chilean government's Ministry of Agriculture, who assist small-scale and *campesino* agricultural development, have pointed in the opposite direction: according to them, the Water Code's impact on poor farmers has been mainly negative. Even though these farmers' fundamental water problems predate the current water law and reflect deeper social and economic problems, the new Water Code seems to have harmed them in several ways.[39]

First, the military government did not provide the public with information, advice, or help in adjusting to the new law. Peasants and small farmers often learned about the new rules and procedures for acquiring or regularizing water rights too late to take advantage of them or to adequately protect themselves. Even when poor farmers have known about the procedures, they have rarely been able to use them without legal, financial, and organizational assistance.

Second, poor farmers are generally unable to participate in the water market except as sellers (if they are fortunate enough to have legal title to water rights, which is uncommon). They lack the money or credit needed to buy water rights. Their main hope for access to additional water is to benefit from the increased return flows that may result from improved irrigation efficiency on the part of more prosperous irrigators upstream. However, downstream users have no legal claim to such unused surplus flows, which are therefore an unreliable and insecure source of water.

Third, peasants and small farmers lack the economic resources and social and political influence needed to defend their interests effectively in the current laissez faire regulatory context, in which private bargaining power is crucial. This is a disadvantage in two areas: conflicts over water use and conflicts over regularizing water rights titles.[40]

Since 1990, the governments of the Concertación have implemented several programs aimed at improving the water rights situation of *campesinos* and poor farmers. Different government agencies have worked together to improve these farmers' legal and physical access to water resources. The programs have included subsidizing small-scale irrigation projects, establishing new water users' organizations, and providing legal and technical help for regularizing water rights titles. These programs have had some important local impacts. On the whole, however, the programs have had very limited funds, and projects have often been blocked by *campesinos'* lack of legal title to water rights. (This problem has been especially frustrating where water may be physically but not legally available because existing water rights are unused by their owners; this has been one of the government's arguments in favor of reforming the law to discourage nonuse.) In any case, programs to assist *campesinos* have been low on the government's larger water policy agenda.

My assessment here is not definitive. I have tried to assemble the available evidence, but there is obviously room for debate and for additional research. Empirical research in this area is challenging because peasants and small farmers generally lack legal title to water rights and prefer to avoid the formal legal system, which makes data collection difficult; without documentary records, fieldwork is the only way to gather information. Moreover, as for most other researchers studying Chilean water markets, my work in this area has not been my main priority.

The uncertainty of the assessment, however, is precisely my point here: it is important to highlight the lack of research about the impacts of Chilean water markets on social equity. Without more knowledge about these impacts, any evaluation of the Chilean experience is incomplete. Although it may be hard to get good empirical information, it is the lack of research effort that is remarkable—and puzzling. Part of the explanation may be that the economic analyses carried out so far have been designed to look at economic efficiency more than distributional issues. One can only speculate about why the more vocal proponents of the Chilean model have not examined the issue more closely.[41]

Within Chile, the lack of research in this area also reflects national political and ideological constraints. Focusing on questions of equity would further politicize the water policy debate, and we saw in Chapter 3 that this debate is already highly politicized. In an ideological context so dominated by free-market economics, Chilean neoliberals commonly argue that the initial distribution of property rights and economic resources does not matter as long as the free market is allowed to work without government interference. (This, of course, is one version of the Coase theorem.) Therefore, in water policy as in other areas, government should aim to lower transaction costs rather than redistribute resources to the poor.[42]

The relative lack of attention to social equity in Chilean water policy is a striking contrast with other developing countries that have adopted major reforms of water law and policy during the 1990s. In both Mexico and South Africa, for example, the social impacts of such reforms—particularly the impacts of the economic and promarket aspects of the reforms—have been a major public concern.[43]

MISSING RESEARCH ISSUE 2: RIVER BASIN MANAGEMENT

The other major issue that has been effectively absent from research about Chilean water markets is the institutional framework for managing river basins. This is really a set of overlapping issues, including resource use as well as environmental protection, since the same institutional arrangements are needed to manage river basins, coordinate different water uses, resolve water conflicts, deal with externalities (both environmental and economic), and define and enforce property rights. These tasks of governance and regulation are the essential core of integrated water resources management, as discussed in Chapter 1.

The lack of an adequate framework for river basin management has been widely recognized in Chile. Since the early 1990s it has been a recurring theme in the debates about reforming the Water Code. DGA's National Water Policy and related documents consider river basin management and conflict resolution critical aspects of attaining the goal of integrated water resources management.[44] Similarly, some World Bank publications about Chilean water markets have pointed out the absence of effective river basin institutions and the flaws of existing arrangements for coordinating multiple water uses and resolving water conflicts. These publications describe the problems as pending challenges for the Chilean government or as issues for future study.[45]

Despite the widespread recognition of those problems, however, most analysis has been brief and superficial. People have tended to mention their existence without investigating them. This has been particularly true of the more institutional and political aspects of river basin management.

In part, the failure to investigate the institutional issues reflects the political constraints described in Chapter 3. From 1993 on, the Chilean government dropped proposals for new river basin organizations from its Water Code reform package because of their hostile reception in the earlier round of public debate.[46] The government made the strategic decision to lower the profile of river basin management, which involved complicated issues requiring extensive discussion and negotiation, and to focus on passing the supposedly simpler reforms first. Because the whole reform process has been longer and harder than expected, river basin issues have remained on the back burner indefinitely.

During the 1990s the Chilean government supported several pilot projects in different parts of the country to explore possibilities for new river basin organizations. These projects were funded by the World Bank and the Inter-American Development Bank and were contracted out to consulting firms. The projects have emphasized the more technical aspects of basin management—for example, reforestation to control soil erosion, and communication among government agencies—and have avoided more conflictive or political issues. Hence the projects have pointed out the need for better voluntary cooperation among stakeholders and government agencies or recommended financial incentives used in other countries (particularly France), but they have not called for greater regulatory powers or specified who would wield such powers over whom. These consulting projects have not led to the establishment of any permanent organizations or to other legislative or regulatory changes, and their overall value is questionable.

Institutional arrangements for conflict resolution in particular have been even more rarely examined in Chile. When the current head of DGA looked back to assess the first 20 years' experience of the Water Code, he singled out the existing processes of conflict resolution as a major problem that still awaits attention.[47] Many Chilean water experts seem to hope that problems of conflict resolution will eventually be solved through some future river basin organizations, and so addressing the problems directly has been postponed. In the absence of river basin organizations, the judicial system is the only alternative to voluntary private bargaining. The lack of research in this area is probably due to the intellectual limitations of academic disciplines as much as to political constraints, as I will argue later.

INSTITUTIONS FOR RESOLVING RIVER BASIN CONFLICTS

My own past research remains the only in-depth empirical analysis of the institutional arrangements for coordinating multiple water uses and resolving water conflicts.[48] The 1981 Water Code says very little about these matters, as described in Chapter 2. The military government and its advisers had other priorities when they wrote the law: first, to clarify and stabilize the water rights situation within the agricultural sector, and second, to establish the legal framework to allow the free trading of water rights, both within agriculture and from one economic sector to another. As a result of those priorities, the management of river basins and the resolution of water conflicts depend on the Water Code's general free-market principles and overall institutional framework, not on any specific or detailed legal rules. That overall institutional framework, in turn, depends on the 1980 Constitution.

In other words, resolving water conflicts in Chile depends primarily on voluntary private bargaining among the owners of water rights. According to the Coase theorem, which is frequently invoked by Chilean economists, this approach should lead to economic efficiency (provided that transaction costs are zero and property rights are clearly defined). Neither DGA nor any other government agency has authority to regulate private water uses or to intervene in water conflicts. When private bargaining fails, therefore, the disputants have nowhere to go but the judicial system—that is, to the ordinary civil courts, since Chile does not have a system of specialized administrative courts. The courts may seek DGA's expert opinion in a particular case but have no obligation to defer to it.

The only exceptions are local disputes between irrigators, which are often settled by private water users' organizations, or canal associations. The main function of canal associations is to distribute water diverted from rivers and streams to the members of a shared canal system, according to the members' water rights, and the associations resolve many routine conflicts among their members. If the canal association cannot settle the problem, irrigators may take their disputes to the local courts, and if not satisfied with the decision there, they may appeal to the higher courts. Disputes between irrigators are often characterized by intricate factual and legal details, but in terms of integrated water resources management, they are the simplest kind of water conflicts and have the narrowest scope and implications.

Conflicts involving nonagricultural water uses, in contrast, fall outside the jurisdiction of the canal associations and go directly to the regional appellate courts. The same is true for conflicts between private rights holders and DGA. The decisions of the appellate courts are often appealed to the nation's Supreme Court. In short, the higher courts are the institutions that matter for all broader conflicts about water uses, regulatory authority, and basinwide water management.

The Water Code does recognize the water users' organizations called vigilance committees, which are federations of canal associations along a shared river. The purpose of these committees is to manage the distribution of water from the river to the head of each canal. Although the vigilance committees have a broader geographic scale than canal associations and sometimes include nonagricultural water users as members, the committees were set up to deal with irrigation only; they have not been able to extend their jurisdiction to intersectoral or basinwide water management.

The institutional arrangements described above are determined by the Constitution as well as by the Water Code. As described in Chapter 2, the Constitution as written by the military regime's legal advisers and adopted in 1980 remained in effect after the return to democratic government in 1990. The 1980 Constitution defines a legal *and* economic model that is characterized by broad private property rights and other private economic liberties, tight restrictions on government regulation and government economic activity (the concept known as the subsidiary state), and a judiciary that has greatly increased authority to intervene in economic and regulatory affairs to protect private rights from government interference.[49] The 1981 Water Code is a faithful reflection of the Constitution in its basic institutional structure and political economic vision, and this is particularly apparent in the laissez faire framework for managing river basins.

EXAMPLES OF CONFLICTS AND MALFUNCTIONING INSTITUTIONS

In this section I will summarize several examples of river basin conflicts that illustrate the flaws in the current institutional framework. These examples are drawn from case studies of two river basins in central and south-central Chile: the Maule River and the Bío Bío River.[50] These two rivers are among the most important in Chile in terms of the economic activity affected and the variety of water uses involved, which include irrigation, hydroelectric power generation, urban drinking supplies, industrial uses, and environmental protection. River basins in other parts of the country, though they differ in their climatic and hydrological conditions, water uses, and specific conflicts, are subject to the same general legal and institutional arrangements.

These arrangements have prevented effective management of externalities caused by water transfers from one river basin to another, conflicts between irrigators and electric companies over how to operate dual-purpose reservoirs, and protection of ecosystems and water quality in both building and operating dams and reservoirs. In all these cases, private bargaining has failed because the economic stakes have been high, the legal rules have not been sufficiently clear, and the relative bargaining power of different actors has been unequal. Neither water users' organizations nor DGA has had the power to clarify or enforce property rights. The higher courts have had the power but lack the technical knowledge and the capacity for policy analysis to do an effective job.

Proposed Interbasin Transfer from the Laja River

A major regional conflict was triggered in 1984 when two competing forest products companies applied to DGA for water rights in the Laja River, for water that they planned to transfer northward out of the Laja basin. The Laja River, the largest and northernmost tributary of the Bío Bío River in south-central Chile, begins in the Andes Mountains at Lake Laja, which was converted into Chile's largest storage reservoir in the late 1950s (see Map 2). By 1981 Chile's national electric company, ENDESA, had built three hydroelectric plants at and below the lake's outlet, which at that time generated almost half the power for the national electric grid. The lake itself was managed for both irrigation and hydroelectric purposes, according to the rules of operation in an agreement signed in 1958 by ENDESA and the National Irrigation Directorate (part of the Ministry of Public Works). Below the lake and the hydro plants, the Laja River supplied water for more than 70,000 hectares of irrigated land and still had abundant volume for plunging over the famous Laja Falls and eventually joining the lower stretch of the Bío Bío River.

The Laja River water was clean, since it was not affected by any polluting activities upstream. The Bío Bío River, however, was badly contaminated both upstream and downstream of its confluence with the Laja. Near the confluence the contamination came from large pulp and paper mills, agroindustrial processing plants, and the provincial city of Los Angeles; farther downstream, the Bío Bío passes through the urban and industrial cen-

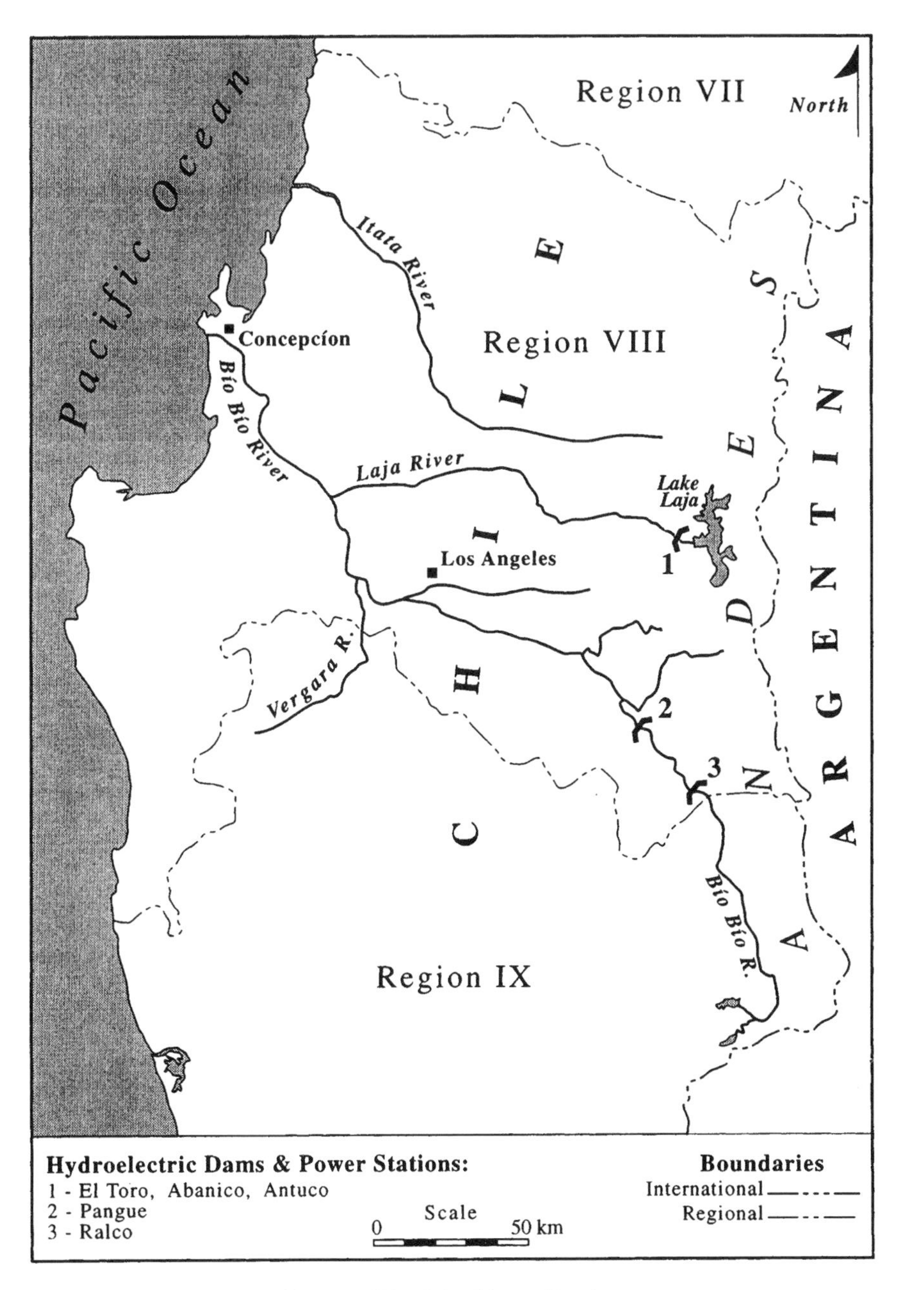

Map 2. Bío Bío River Basin

ter of Concepción. Located at the river's mouth, Concepción is the regional capital, a metropolitan area with 750,000 inhabitants whose drinking water comes from the river. It is also the site of some of the country's largest steel and petrochemical industries, which both used and polluted the water in the lowest reach of the river. In short, the Laja River performed a critical function by providing large volumes of clean water to dilute the pollution in the lower Bío Bío.

The forest companies' proposals to transfer water northward out of the Laja basin meant transferring it out of the Bío Bío basin as well. Both companies proposed to build new hydroelectric plants in the next river basin to the north. Because there was only enough water available for one of the proposed projects, however, and because the applications had been filed within a month of each other, DGA was required by the Water Code to auction off the water rights to the highest bidder. That put DGA at the center of attention and debate.

The proposed interbasin transfer was strongly opposed by many downstream water users in both the Laja and the Bío Bío River basins. Opponents included local irrigators and canal associations; regional business interests, university experts, and professional associations in Concepción; and the regional offices of two national government agencies, one responsible for sewage and sanitation works and the other for promoting tourism. These opponents protested that the transfer would increase the concentration of pollution in the lower Bío Bío and would dry up the Laja Falls. Irrigators protested the impact on the security of their water supplies: in theory their water rights would continue to be respected, but they feared being trapped between two big nonagricultural water users over which they had no influence—the new out-of-basin transfer and ENDESA's existing hydroelectric plants upstream.

DGA considered the protests and imposed certain conditions on the new water rights but decided to go ahead with the auction in late 1985. The conditions were that the rights could be used only during the rainy season and that a minimum flow would have to be maintained at the Laja Falls downstream. The projects' opponents considered those conditions insufficient and doubted their enforceability once the new hydro plant was in operation. Finally, a coalition of business and university interests in Concepción filed a lawsuit against DGA, arguing that the auction would violate their constitutional right "to live in an environment free of contamination."

DGA defended itself by arguing that it recognized the problem but had no choice about granting the water rights. According to a strict interpretation of the letter of the law, the agency had no regulatory authority over pollution control and no discretionary power to call off the auction, even in view of the likely impacts on water quality. Half a year later, both the Appellate Court of Santiago and the Supreme Court rejected the lawsuit and fully supported DGA's position. The courts' reasoning was entirely formalistic, and their reluctance to get involved was evident. They ruled that DGA's refusal to go beyond its explicit legal authority was justified, and that since there were no procedural errors, the judges themselves had no obligation to examine the substantive constitutional issues that the plaintiffs had raised.

Nevertheless, the lawsuit succeeded in delaying the auction long enough for the plaintiffs to negotiate a political solution. The Minister of Public Works (a military general) intervened from Santiago to bring the parties together, and the two forest companies eventually agreed to drop the proposed projects and withdraw their application for water rights. The companies acted mainly in the interests of good public relations, since both were heavily invested in the regional economy and their opponents in this matter were politically influential.

The political context and bargaining environment of this conflict were different from the examples I discuss next, since the military government was still in power. Nonetheless, the institutional framework is the same today, and this case illustrates a pattern of institutional behavior that has been common in other examples of river basin conflicts: a preference for narrow and self-protective legalism on the part of both DGA and the courts, which combine to avoid difficult problems rather than face them.[51]

Irrigation versus Hydroelectricity: Consumptive versus Nonconsumptive Water Rights

The one area in which the 1981 Water Code specifically addressed multiple water uses was in the creation of a new kind of property right: "nonconsumptive" water rights. A nonconsumptive right allows its owner to divert water from a stream or river and use that water to generate hydroelectric power, provided that the water is then returned unaltered to its original channel (though not necessarily to the original point of diversion). The goal was to encourage hydroelectric development in the upper parts of river basins—the mountains and foothills—without harming the existing rights of irrigators downstream in the agricultural valleys. By the time the Water

Code was enacted, most of the surface waters in the agricultural areas of central and northern Chile had already been fully allocated as "consumptive" water rights. Hence the invention of nonconsumptive rights aimed at intensifying the uses of water resources without having to compensate the owners of existing rights.

Unfortunately, the Water Code's rules defining the relationship between consumptive and nonconsumptive rights are brief and ambiguous, and coordinating them has been much harder than expected. These problems have been most fully developed in the Maule River basin, as summarized below.

The law is not entirely clear about whether either right has priority in case of conflict. Several provisions suggest that nonconsumptive rights are subordinate to consumptive rights—particularly Article 14, which defines nonconsumptive rights and requires that "the extraction or restitution of waters shall not damage the rights of third parties to the same waters, in terms of their quantity, quality, substance, opportunity of use, and other details."[52] ("Opportunity of use" is the critical phrase, as illustrated in the conflicts described below.) On the other hand, the code does not generally recognize any order of preference among different kinds of water uses when, for example, people apply to DGA for new rights, since the law's basic principle is that priorities should be determined by private owners and the free market rather than by legislation. Besides the article quoted just above, the code does not offer additional rules about how to exercise nonconsumptive rights. DGA determines any additional rules individually for each nonconsumptive right at the time that it is formally granted.[53]

Another problem is that the rules for decisionmaking within vigilance committees are biased in favor of nonconsumptive rights. I mentioned above that the main function of these committees is to distribute water from rivers to different canals according to their water rights, and as a result the committees have historically been composed entirely of irrigators. Important decisions within the committees are made by majority vote of the members, who cast votes in proportion to their water rights. The current Water Code, however, says that the owners of nonconsumptive rights are also members of vigilance committees. The code's drafters apparently failed to notice that nonconsumptive rights will outnumber consumptive rights in any basin with more than one hydroelectric plant, simply because there is a separate right for each nonconsumptive use of the same water.[54] In practice, the vigilance committees have remained dominated by irrigators who in self-defense tend not to invite the power companies to their

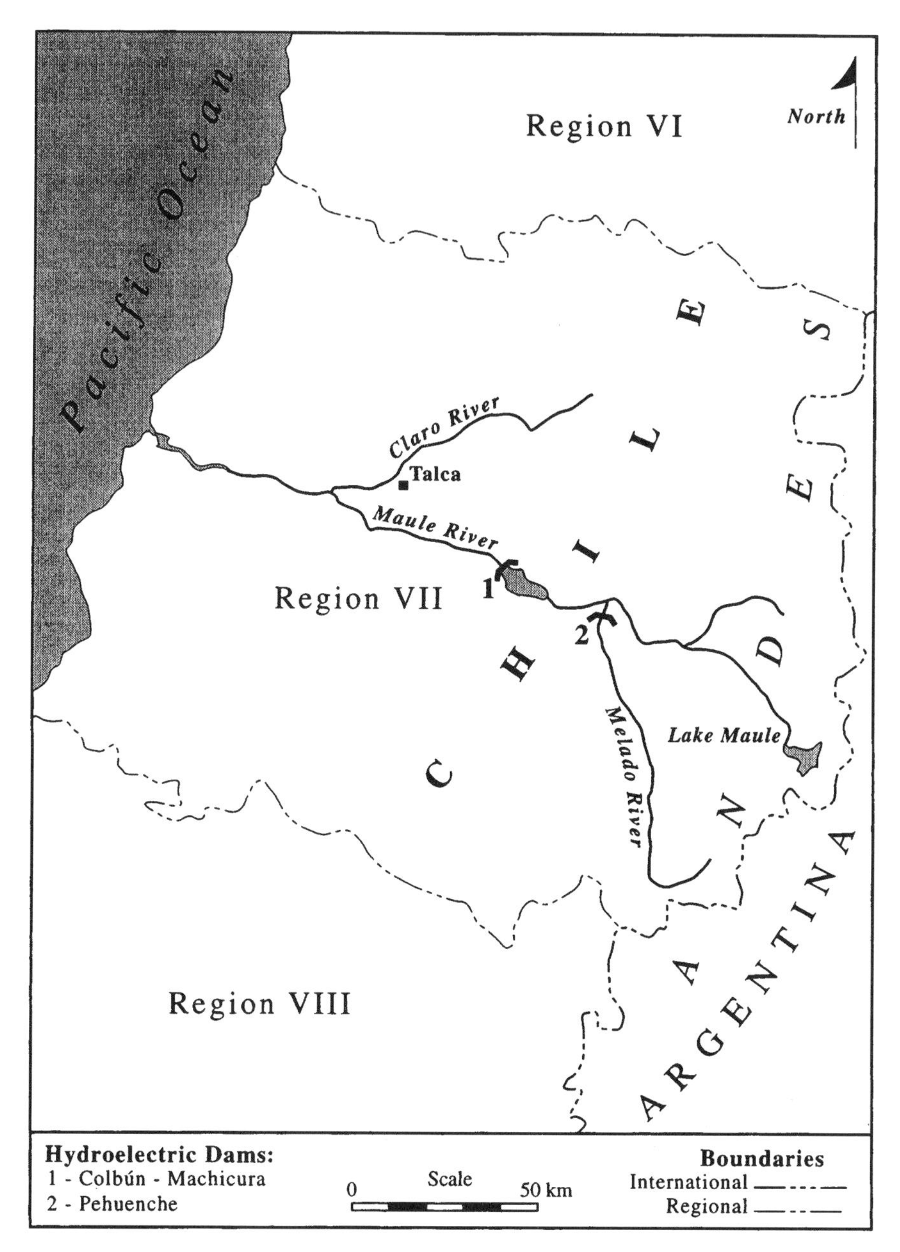

Map 3. Maule River Basin

meetings; the power companies, in turn, deny the committees' authority over their water use.

The conflicts between consumptive and nonconsumptive rights have been sharpest in the management of multipurpose dams and reservoirs. Power companies and farmers have conflicting seasonal demands for water in Chile: power companies want to store water during the summer to meet high national electricity demand in winter, while farmers want to store water during the rainy winter for use in the summer growing season. The will to cooperate is undermined by a single-use mentality that is deeply rooted on both sides.

The Water Code's ambiguity about the relationship between the two water uses also reflects the situation when the law was drafted: ENDESA, its water rights, and all significant dams and reservoirs were then owned by the national government; their management was under government control, and there was little need to spell out regulatory precautions. Unfortunately, the military government did not modify the Water Code when it privatized ENDESA in the late 1980s, and the company's water rights were included in its assets. These conflicts have been further complicated by the fact that the management of Chile's two most important dual-purpose reservoirs is governed by legal agreements that predate the 1981 Water Code and are still in force. Those reservoirs are Lake Laja, described above, and Lake Maule, described below.[55]

In short, the inadequacy of the Water Code in this area has led to serious legal and policy problems about how to interpret and enforce the relationship between consumptive and nonconsumptive water rights. Responding to these problems has been a major test of the current institutional framework. Moreover, these problems are distinct from the issues of nonconsumptive rights that have dominated political debates over reforming the Water Code, as discussed in Chapter 3. Those debates have focused on issues of speculation, hoarding, and monopoly power, not on issues of managing dams or river basins.

Maule River Basin

The Maule River basin has been the site of the most complex conflicts between consumptive and nonconsumptive water rights. This basin has about 200,000 irrigated hectares and is part of Chile's traditional agricultural heartland, 250 kilometers south of Santiago (see Map 3). The main section of the Maule River has more than 30 canals and 10,000 irrigators, who are members of the Maule River Vigilance Committee. Hydroelectric development on a moderate scale began in the upper basin in the 1940s, when ENDESA and the National Irrigation Directorate signed an agreement to convert a mountain lake, Lake Maule, into a dual-purpose storage reservoir (as in the case of Lake Laja, but much smaller). Large-scale hydro development began in the 1980s, after passage of the current Water Code.

ENDESA built the primary dam, called Colbún, where the Maule River leaves the Andean foothills and enters the valley floor. The reservoir has the capacity to store approximately 20% to 30% of the river's average annual flow. In 1983, DGA granted ENDESA the basin's first nonconsumptive rights, which stated that existing consumptive rights in the river could not be harmed and specified a month-by-month schedule of guaranteed minimum flows that Colbún would release to irrigators downstream. Those flows were highest during the summer irrigating season. Colbún's rights also stated that the dam would become a member of the river's vigilance committee. Colbún began operation in 1985, producing about 25% of the power used by the national transmission grid, and caused no serious problems for irrigators.[56]

The problems were triggered by ENDESA's second large dam in the Maule basin, called Pehuenche, which was built upstream from Colbún. DGA granted ENDESA nonconsumptive rights for Pehuenche in 1984. As with Colbún, those rights stated that Pehuenche would become a member of the vigilance committee. Pehuenche Dam differed from Colbún in important ways, however. First, Pehuenche has a much smaller reservoir and little storage capacity: it is essentially is a run-of-the-river dam, although its generating capacity is similar to that of Colbún. Second, the military government had privatized ENDESA by the time Pehuenche was built, and Pehuenche was a subsidiary of ENDESA. Colbún, in contrast, had been parceled off as a separate company and was still publicly owned, to provide competition for ENDESA. Colbún retained its water rights and its legal obligations to release guaranteed flows to irrigators.

When Pehuenche Dam was finished in November 1990, the company closed the dam's gates to fill the empty reservoir and thus be able to generate power a few months later. This was late spring in a drought year—a time of high water demand for irrigation—and downstream farmers immediately complained to the vigilance committee about the interruption of the river's flow. The committee asked Pehuenche to reopen the dam's gates and let the water pass. Pehuenche refused and the committee went to the regional appellate court, in the nearby city of Talca.

The legal battles that followed dragged on for years and involved intricate details and maneuvers that cannot be recounted here. Instead I will try to summarize the critical issues, arguments, and events.[57] Keep in mind that when I refer to Pehuenche, ENDESA is the company pulling the strings.

Both the Maule River Vigilance Committee and DGA argued that Pehuenche was violating the irrigators' property rights, first by filling the new reservoir and later by regulating the river's flow to suit the needs of power generation. They argued that the Water Code clearly said that nonconsumptive rights could not interfere with preexisting consumptive rights (see Article 14 as quoted above). Hence Pehuenche would have to negotiate with the irrigators from that basis, and either modify its operation or offer compensation. (Since Pehuenche's reservoir capacity was limited, it could regulate the river flow on a daily or weekly basis only, according to peak electricity demand.)

Partway through this dispute, Colbún joined the vigilance committee and DGA. Colbún had to comply with its own obligations to release water to the irrigators, regardless of what Pehuenche did upstream, and was thus paying for Pehuenche's plan of operations. Pehuenche could generate power whenever the prices for electricity were highest, an advantage that Colbún did not have. Colbún argued that Pehuenche should pay compensation for Colbún's services in mitigating the impacts of the upstream company's actions.

ENDESA and Pehuenche counterattacked with the opposite interpretation. The power company argued that the Water Code did not establish any priority between consumptive and nonconsumptive rights, and furthermore that nonconsumptive rights implicitly included the right to fill reservoirs and temporarily regulate river flows; otherwise the intended hydroelectric development would not be possible. ENDESA also sued DGA, arguing that the agency's actions in defense of the irrigators violated ENDESA's property rights and that DGA had no legal authority to tell ENDESA how to use its water rights.

DGA agreed that it had no such authority—it could only offer its legal interpretation to the courts. But here DGA stood firm, arguing that nonconsumptive rights did not include the right to freely alter river flows. DGA's position was that ENDESA had to negotiate its regulation of river flow with the other owners of water rights.

The appellate courts in Talca and Santiago initially agreed with the interpretation of DGA and the vigilance committee and ordered Pehuenche to release water to the irrigators. But those decisions were appealed to the nation's Supreme Court, and the Supreme Court derailed and confused the whole process on several occasions. Although the appellate courts tried to grapple directly with the substance of the legal issues, the Supreme Court's tendency was to avoid the issues as much as possible—in ways that ended up favoring ENDESA.

The Supreme Court's first ruling in this conflict was to remove from the record a long analysis by the Appellate Court of Santiago. The judges on that appellate court had tried to reason their way through the question of how to interpret the mutual obligations of consumptive and nonconsumptive rights. They concluded that DGA's and the vigilance committee's interpretations of the law were more reasonable than ENDESA's. The Supreme Court, however, refused to address that question, although it agreed that in the particular situation at hand the irrigators' property rights were threatened. Instead the Supreme Court omitted the appellate court's reasoning and referred the substantive conflict to DGA to resolve, despite the agency's lack of authority to do so.

The Supreme Court's next ruling, a few months later, again set aside an appellate court decision in favor of the irrigators, and again refused to decide the substantive issue. Instead, the Supreme Court persuaded ENDESA, Pehuenche, and the vigilance committee to simply drop their various suits and countersuits without any resolution. At that point the Pehuenche reservoir had already been filled (thanks in part to a transfer of water from Lake Maule upstream, approved by ENDESA and the National Irrigation Directorate) and the current irrigating season was over. The vigilance committee had effectively lost the first round.

When the same conflict recurred the following summer, the Appellate Court of Talca took its cue from the Supreme Court and refused to address the issues. Instead, it insisted that Pehuenche and the irrigators should enter private arbitration. The arbitration process went on slowly for years and eventually narrowed to just one question: whether Pehuenche owed compensation to Colbún. The vigilance committee withdrew from the process, and in the end the arbitration resolved nothing about the relationship between consumptive and nonconsumptive rights. That larger issue was finally addressed in the landmark case of the Pangue Dam.

Pangue Dam on the Bío Bío River

The turning point in the legal conflict between consumptive and nonconsumptive water rights was a case involving ENDESA's Pangue Dam on the upper Bío Bío River (see Map 2). This dam has been nationally and internationally famous as a controversy about environmental impacts and the rights of indigenous peoples; its water rights issues have been much less known. Nonetheless, a landmark decision by the Supreme Court focused directly on water rights and showed both the court and DGA in a particularly unfavorable light.

The case turned on many of the same issues as those presented by Pehuenche Dam in the Maule basin. Pangue Dam, built by ENDESA, would generate a lot of power but would have a fairly small reservoir.[58] In 1993 a coalition of environmental and indigenous organizations and downstream irrigators and canal associations sued ENDESA to prevent Pangue's completion (construction had already begun). The heart of their legal argument was familiar: Pangue's nonconsumptive water rights did not include the right to alter the river flow without regard to the impacts on downstream rights holders. ENDESA responded by repeating all of its arguments to the contrary.

The Appellate Court of Concepción ruled against ENDESA, just as the Appellate Court of Santiago had done when first confronted with the problem, and with the same legal reasoning. The court ordered ENDESA to stop construction until it had reached an agreement with other water rights owners about how the dam would be managed. This decision triggered six weeks of heated national political debate before the Supreme Court reversed it.

In reversing the appellate court's decision, the Supreme Court finally ruled on the substantive issues at stake. The court abandoned its earlier protection of the property rights of irrigators, dismissed their concerns as exaggerated and premature, and accepted all of ENDESA's arguments in favor of nonconsumptive rights. Pangue could resume construction, fill the completed reservoir, and develop its plan of operations without negotiating with other water users. Anyone whose rights were affected in the future could sue later.

The Supreme Court relied in part on a report by DGA in which the agency reversed its own position in the Maule conflict. This time DGA said that the Pangue Reservoir did not necessarily threaten downstream water rights, although its operation should be monitored in case of future problems. It is hard to interpret this DGA report as anything other than a response to political pressure from higher levels of the government, so as not to hinder electricity generation.

In the months following the Supreme Court's Pangue decision, the appellate courts in Santiago and Talca applied the same ruling to several problems of integrated water resources management. The results have been discouraging. Private bargaining has failed to resolve these conflicts. Both DGA and the higher courts have demonstrated their preference for a strict interpretation of the law and highly legalistic behavior. In routine matters this combination has worked reasonably well: DGA has been careful to observe the limits of its authority, and the courts have generally deferred to the agency's technical expertise while occasionally correcting evident administrative errors. With more complex problems, however—problems that involve technical details, difficult regulatory issues, and high economic and political stakes—the combination has worked quite poorly.[60]

Faced with difficult problems of public policy and ambiguous legal provisions, such as the exercise of nonconsumptive water rights or the impacts of interbasin water transfers, DGA has tended to adopt a narrow legalistic position to protect itself from political criticism and constitutional challenge. The agency has been so careful not to exceed its explicit regulatory duties that it has nearly always erred on the side of passive rather than assertive behavior.[61] This kind of regulatory behavior is precisely what the military government and its neoliberal advisers intended in the 1980 Constitution, and it conforms to their concept of the subsidiary state.

Unfortunately, however, the judiciary has frequently avoided its own expanded responsibilities within this institutional framework. The higher courts in particular have often found formal or procedural reasons to avoid ruling on the substance of difficult policy issues, whether because the issues are technically hard to understand or because they involve political as well as legal matters. When the courts have decided such issues, they have often done so on the basis of narrow or superficial analyses of the legal principles and the range of public and private interests involved. The Supreme Court has led the way in establishing this pattern, and in Chile the Supreme Court exerts strong and centralized control over the national judicial system.

Those institutional arrangements have left a partial vacuum in regulatory decisionmaking, involving precisely the kind of difficult issues that require not only technical expertise but also political judgment. In this partial vacuum, the more powerful actors can generally do as they please. The most direct way to resolve some of these issues would be to clarify the rules of the game through new legislation. Indeed, for Chile's current constitutional and regulatory framework to work effectively, there must be frequent intervention by the lawmaking branches of government. Otherwise, when administrative agencies lack discretionary authority and courts have broad

pending cases in the Maule basin. As far as the judicial system w
cerned, this question of legal interpretation had been settled.

From the standpoint of public policymaking, however, the S
Court's performance and final decision were seriously flawed—
dictable, formalistic, and superficial. The extreme positions taken l
ENDESA and the irrigators—that one kind of water right shou
absolute supremacy over the other—were understandable in a priva
conflict. From the perspective of integrated water resources manag
however, a more reasonable outcome would have been some kind
promise, or at least the encouragement of some process for more b
negotiation among competing water uses. The Supreme Court unde
these possible outcomes by sending clear signals to the appellate
that attempts to wrestle with the substantive issues were not welco
when the Supreme Court finally addressed those issues itself, it ad
narrow and one-sided interpretation of the law, imposing new limi
on vested property rights without requiring the owners of the new
more profitable kind of water rights to concede anything in ex
(either financially or in principle). This was a major transfer of wealt
irrigators to electric companies, and a significant redefinition of p
rights, on the basis of legal reasoning of dubious quality.

It is ironic that Chilean neoliberals so often refer to the Coase th
as one of the core principles of the Chilean water law model. U.S.
mist Yoram Barzel, an expert in property rights trained at the Unive
Chicago, has pointed out that the river basin problems described
have a simple theoretical explanation: the assumptions of the Coase
rem have not been met. In the first place, property rights are not
defined. The Water Code may define both consumptive and noncon
tive water rights as private, exclusive, and tradable, but that is only a
ning. Those definitions are not nearly precise enough to handle the
and complexity of the situations in which those water uses interact.
second place, transaction costs are obviously significant. In the cont
multiple water uses and river basin management, it is an illusion to
that transaction costs could ever be close enough to zero for the Coas
orem to work.[59]

EVALUATING THE INSTITUTIONAL FRAMEWORK

The river basin conflicts summarized above indicate how the Chilean
and institutional framework has responded to some of the fundam

powers to review administrative actions but are unable or unwilling to address policy issues, paralysis is the likely outcome.

In many areas of public policy, however, such regular legislative action has been prevented by political deadlock and eclipsed by other government priorities. Water policy is not the only example. There is no reason to expect these dynamics to change in the foreseeable future. We might expect that the growing intensity and complexity of water conflicts will eventually force some legislative action, but the danger is that such action may be driven by immediate crisis rather than by careful deliberation.

In this legal and institutional context, creating new river basin organizations as they are usually discussed in Chile would almost certainly be ineffective. Such organizations are commonly described as being based on broad participation and coordination of both private and public actors, but with little centralized or top-down authority. Any effort to create organizations with regulatory teeth and adequate budgets would run into opposition from both the private and the public sectors. The organizations would be handicapped by the strong private economic rights of water users and by constitutional restrictions on regulatory power. Their actions would be subject to judicial review, just as other government agencies are. And their mission and authority would overlap with those of other government agencies. One proposed alternative—building on the existing vigilance committees—does nothing to avoid these substantive problems.

The solution would be a legal agreement that explicitly addressed the constitutional obstacles and was supported by political majorities large enough to overcome constitutional objections. We saw in Chapter 3 that such a degree of political consensus in Chile is extremely unlikely, however. As a result, the discussion of river basin organizations remains stuck between vague rhetoric and weak proposals that would make no difference in the difficult problems of water management.

Given the obvious importance of these institutional issues to water management in Chile, it is striking that there has been so little research about them, by either Chileans or foreigners. I first published my argument about the inadequacy of Chilean courts for dealing with water conflicts in 1993. Since then the argument has often been repeated in Chile, both in government documents and in academic papers, and it has also been cited in some of the World Bank's publications—indeed, the argument seems to have become part of the conventional wisdom about Chilean water management. In later publications I continued to emphasize the central and problematic role that the courts play in Chile's regulatory framework.[62]

Other researchers, however, have not pursued the issues. There have been no additional studies of judicial performance and decisionmaking in water conflicts, or of patterns in judicial reasoning and interpretation, or of the relationship between the courts and DGA, or of nonjudicial alternatives to conflict resolution. Even the specific legal problems raised by the relationship between consumptive and nonconsumptive water rights, despite their notoriety, have remained untouched by academic researchers and absent from the government's proposed Water Code reforms.

Although the government's neglect of these issues is due to their political sensitiveness, the lack of research also has a deeper explanation: the intellectual limitations of academic and professional disciplines. Both in Chile and elsewhere in Latin America, interdisciplinary studies are rare. The subject matter and analytical methods of different disciplines tend to be narrowly defined and the boundaries between disciplines strictly policed.

The most important disciplines in the field of water resources management in Chile are engineering, economics, and law. Engineers are the most numerous professionals in this field, and their work experience often makes them knowledgeable about water management. Nonetheless, the technical nature of engineering means that there is little room for the social and political sciences or for legal, institutional, and policy analysis.

Economists have not been much better when it comes to research in this area. Resolving conflicts and managing river basins are essentially matters of legal, institutional, and political arrangements. Virtually all the economic research about Chilean water markets, however, has been done by economists trained in the orthodox neoclassical tradition. They have lacked the background, analytical methods, and often the interest needed for studying institutional issues, except as the background for water rights transactions. At best, they recognize the broader problems without examining them further.

For example, in Guillermo Donoso's survey of Chilean water markets, discussed earlier in this chapter, he dismisses the conflicts between consumptive and nonconsumptive water rights as "not a very relevant problem, because they have only occurred in a few basins."[63] That is the extent of his analysis of the topic. Yet those basins are among the most important basins in Chile, and the problem will recur in other basins as hydroelectric development continues. More importantly, Donoso misses the broader significance of the problem as an indication of the flaws of the institutional framework for coordinating water uses and resolving river basin conflicts.

The point here is not whether his opinion is correct, but rather, that he does not examine the issue itself more thoroughly.

For their part, lawyers in Chile have been trained in a legal tradition that is highly formalistic and focused on the letter and the abstract logic of the law in deliberate isolation from its social and political context. This is the "law on the books" approach, as opposed to the interdisciplinary "law in action" field of sociolegal studies—a field that barely exists in Latin America. In Chile the narrow and semantic approach to law so dominates that there has been virtually no research about the legal aspects of water rights or river basin management in practice. Legal scholars as well as judges have restricted themselves to "pure" legal analyses, which have contributed little to practical understanding or public policy debates about water management issues.[64]

Finally, research in the ecological and environmental sciences has had little influence in river basin management. DGA and other government agencies sometimes contract such research from universities and may take it into account in their own decisionmaking. But the agencies' lack of regulatory power means that the environmental information has had limited impact on water uses. Environmental protection in Chile is still weak in relation to property rights and the pressures for economic growth.

EMERGING ISSUES IN CHILEAN WATER POLICY

Several new water policy issues have emerged in Chile since the mid 1990s, in addition to the issues discussed so far. They reflect the continuing evolution of economic and political conditions in Chile, including changes in public opinion, media coverage, the balance of political forces, and the concerns of NGOs and other interest groups.

My purpose here is simply to mention these new issues briefly to convey a sense of the growing complexity of the field of Chilean water policy and water resources management. All these issues have increased pressure on the existing legal and institutional framework and therefore have highlighted the need to improve institutional and regulatory capacity. Although the emerging issues are closely related to water markets, for the most part they fall outside the scope of existing research about those markets and thus offer important areas for future research.[65]

Without attempting to list them in order of importance, the emerging issues include the following:

- *The continuing uncertainty, confusion, and complexity of legal title to water rights in many parts of the country.* The government has made gradual progress toward creating and maintaining a comprehensive, up-to-date national register of water rights titles, but this process is still in its early stages. There is broad political consensus that modernizing the legal and hydrological information about water rights will be a vital step in improving the performance of water markets and water resources management.
- *The privatization of urban water supply companies in the more populous regions of the country, particularly the metropolitan areas of Santiago, Valparaiso, and Concepción.* The government's decision to maintain ownership of the companies' water rights (in contrast to the military government's decision to privatize ENDESA's water rights along with the company's other assets) has been controversial. It remains to be seen how the private companies' water use and management will fit into the general context of river basin management and coordination with other water uses.
- *The development, use, and regulation of groundwater.* Groundwater was formerly a little-used resource but in recent years has been in great demand, especially in the northern desert regions, where surface water is extremely scarce. Large mining projects, growing cities, and expanding cultivation of high-value agricultural crops have greatly increased the demand for water in arid areas, but DGA has refused to grant all the new groundwater rights that have been requested, and as a result the agency has been entangled in conflicts with private sector interest groups and with the government's Office of the Controller General. DGA has argued that hydrological data do not prove the physical availability of sufficient groundwater.
- *The water rights and needs of Chile's indigenous communities, particularly in the northern desert and mountain areas.* These communities' dependence on crops and livestock has been threatened by competing demands for water from mining projects and the growing cities on the northern coast. These conflicts have been further complicated by indigenous peoples' distinct cultural views of the importance of water, which are often incompatible with the commodity logic of the current Water Code. In southern Chile conflicts between indigenous communities and ENDESA continue over the construction of hydroelectric dams in the upper Bío Bío River basin: after the Pangue Dam was finished, ENDESA began the Ralco project, which is scheduled for completion in 2004.

- *The efforts of the government to promote private concessions for building and operating new water-related infrastructure, particularly reservoirs for irrigation.* The Ministry of Public Works has adapted the government's system for granting private concessions for other kinds of infrastructure development, such as highways and port facilities. With reservoirs this has raised two questions that have not yet been answered: first, under what conditions can such projects be profitable without large government subsidies, and second, how will the management of these reservoirs fit into the existing institutional framework for river basin management, including the coordination of multiple water uses and resolution of water conflicts?
- *The growing need for environmental protection of water resources and river ecosystems.* DGA has attempted to establish minimum in-stream flows for ecological purposes, which is a difficult task in a country where most water rights have already been allocated and where regulatory authority is weak. The national environmental legislation passed in 1994 has provided some support for protecting in-stream flows, mainly in the process of evaluating the environmental impacts of proposed dams. Apart from in-stream flow protection, the level of water pollution has begun to improve in some urban areas, thanks to the construction of new sewage treatment plants. These new plants, built as part of the privatization of water companies, will raise the prices paid by urban water consumers. A great deal remains to be done in all these areas in a country where the economy is based on the exploitation and export of diverse natural resources.
- *The long-standing and highly controversial debates about legal reforms in electricity regulation.* Although their final shape is not clear, these reforms have potentially major impacts on river basin management because of the country's heavy reliance on hydroelectric power. Moreover, the institutional arrangements for conflict resolution in the electricity sector share the problems discussed above in relation to water rights—namely the weakness of regulatory agencies and the flaws in the judicial system.

CHAPTER 5

CONCLUSIONS AND LESSONS ABOUT THE CHILEAN EXPERIENCE

Chile has now had more than 20 years of experience with the world's most pure free-market water law, and this experience offers valuable lessons for both national and international water policy reforms. The Chilean model of water rights and water resources management is important because it is such a paradigmatic example of the free-market approach to water law and economics, and thus the Chilean experience is relevant to the broader international water policy issues described at the beginning of this book.

As discussed in Chapter 1, since the early 1990s there has been increasingly urgent international debate about a global water crisis. Growing demand and competition for water resources worldwide have caused increasing scarcity and conflict. As these problems have become worse and more obvious, they have led to a broad international consensus about the need for major reforms in how water is used and managed—reforms that are often described as pointing toward more integrated water resources management (IWRM). IWRM aims to be holistic, comprehensive, interdisciplinary, and sustainable over the long term. The goal is to incorporate all three of the elements that are generally considered to make up sustainable development: economic efficiency and growth, social equity, and environmental protection. Addressing all these issues in practice, of course, is extremely difficult. Growing recognition of this difficulty has inspired the recent international focus on water governance, with its increased concern for the social, political, and institutional aspects of water management, rather than the more technical aspects that have traditionally dominated the field.

One of the core principles of IWRM is that water should be recognized as an "economic good." But what does this phrase mean, and what are the policy implications? How do they relate to the broader goals of IWRM and better water governance? I showed in Chapter 1 that the range of economic perspectives on these questions reflects different intellectual, disciplinary, and political approaches to economic analysis. I described this range as a spectrum from "narrow" to "broad" economic perspectives, where narrow refers to more technical and mathematical neoclassical analyses (which may or may not have an ideological preference for free markets), and broader perspectives draw more on history and other social sciences and focus more attention on institutional issues.

For the purposes of this book, the crucial distinction between the different economic perspectives is how they address the institutional context and foundations of markets—in simple terms, the rules of the game. Institutions here mainly consist of legal and political systems, although they may also include other social norms. As I explained in Chapter 1, I am not suggesting that those with narrow economic perspectives ignore or deny the importance of such institutional arrangements. What I am saying instead is that their analysis of such arrangements generally does not go very deep and is determined by the assumptions of neoclassical theory about the laws, institutions, and government required for markets to work. For example, the analysis depends on clearly defined property rights and enforcement of contracts, without delving further into the social and political complexities implicit in such assumptions.

With respect to the concept of water as an economic good, there are two general economic perspectives. One is that water should be treated as a scarce resource; the other is that water should be a private and marketable commodity. Both positions agree that allocating water involves difficult choices and trade-offs, for which economic incentives can be powerful tools. But the two positions have quite different implications for the role of markets and for legal, institutional, and regulatory arrangements, as the example of Chile illustrates clearly.

The institutional arrangements that are critical to IWRM are those required for defining and enforcing property rights, coordinating different water uses within river basins, internalizing environmental and economic externalities, resolving conflicts, and monitoring compliance. By their nature these processes are all interrelated. Moreover, they have unavoidable social and political aspects, since they allocate costs and benefits among different people and determine who gains and who loses. Because different economic perspectives shape the design of these institutional arrangements, the stakes are high.

In short, the essential question posed in this book is whether a narrow and free-market approach to economics is compatible with IWRM. The Chilean experience, I argue, shows that the answer is no. Although this approach has some economic benefits, its institutional consequences have led to serious structural problems in management and regulation. The advantages of the Chilean model of water law and water markets come with significant disadvantages, and hence the model can be considered successful (or an example of good practice) only by criteria that are too limited. This explains the book's title and the allusion to Odysseus: the Chilean model has been a song of the sirens to water policy reformers in other countries because it seems so attractive that people fail to see its dangers.

In this chapter I will summarize first the conclusions and lessons that are specific to Chile, and then the conclusions and lessons for other countries and the international arena.

CHILE: LOOKING BACK FROM 2004

If we review the more than 20 years since the passage of the 1981 Water Code, and particularly the period since Chile's return to democratic government in 1990, two trends stand out.

First, there has been a great deal of political and policy debate about water rights and water markets in Chile, but there has been much less academic or empirical research—that is, research attempting to be impartial rather than tied to particular political positions. Much of the academic research has been done by North Americans, and much of the policy research has been done by people in international organizations. The relative lack of empirical research by Chileans is both a cause and an effect of the highly politicized nature of the domestic debate about water policy.

Second, two related issues, or sets of issues, have dominated discussions in both policy and academic arenas.

- What should be the basic legal rules defining property rights to water? How have the current rules, which are extremely laissez faire, affected the economic incentives for water use? Are there better alternatives?
- How have water markets worked in practice? What have been their strengths and weaknesses, and how might they be improved by legislative or regulatory reforms?

Both sets of issues have been hotly contested in Chile. Outside the country, in contrast, nearly all attention has been directed to the second set, Chilean water markets. I examined these issues in detail in Chapters 3 and 4, and I will address them here in reverse order: first the empirical results of water markets, and next the political debate about reforming the Water Code.

EMPIRICAL RESULTS OF THE 1981 WATER CODE

We can assess the empirical results of the current Water Code in two ways: (1) by comparing them with the law's original objectives; and (2) by comparing them with the issues considered critical for integrated water resources management.[1] The law's original objectives, of course, were determined by the strong political and ideological views of the Chilean military government and its civilian advisers, as discussed in Chapter 2. Many people both inside and outside Chile do not share those views. In comparing the law's results with those objectives, therefore, I do not mean to imply that the objectives themselves were necessarily correct.

Comparison with Original Objectives of the Water Code

It is important to recall a few general points from Chapter 2 concerning the Water Code's political background and legislative history. When the Water Code was being drafted in the late 1970s and early 1980s, the primary concerns were irrigation and agriculture. Relatively little attention was given to nonagricultural water issues, other than to assume that free markets would reallocate some agricultural water supplies to nonagricultural uses. As far as other issues were concerned, free-market economic theory was simply applied to water management without much effort to adapt it to the specific characteristics of water resources. The best example of this is the laissez faire approach to coordinating water uses within river basins. Another example is the inadequate definition of the rules governing the new category of nonconsumptive water rights. Finally, recall that issues of water quality and environmental protection were left out of the Water Code almost entirely.

It is also important to keep in mind that some of the fundamental elements of the Chilean water law model have been determined by the 1980 Constitution more than by the 1981 Water Code. One such element is the constitutional provision that declares water rights to be private property, which enjoys very strong protection from government intervention. A second such element is the basic structure of the overall institutional and regulatory framework, particularly the limited power of the government administrative agency, the Dirección General de Aguas (DGA), and the strong and broad authority of the judiciary. This framework is what determines the resolution of water conflicts and related institutional functions.

The Water Code, together with the Constitution, has been fairly effective in achieving several of its original priorities, mainly those having to do with strengthening private property rights.

- The legal security of private property rights has encouraged private investment in water use and infrastructure. The level of this investment has varied in different parts of the country; it has allowed new mining development, especially in northern Chile, and the planting of high-value fruits and vegetables for export.
- The counterreform in agrarian land tenure has been consolidated.
- Government regulation of water use and water management has been tightly restricted.
- The freedom to trade water rights has allowed reallocation of water resources in certain circumstances and geographic areas.
- The autonomy of private canal users' associations from government has been affirmed, which in some cases has encouraged these organizations to improve their administrative and technical capacity; however, these organizations operate only within the agricultural sector and rarely include nonagricultural water users.
- The creation of nonconsumptive water rights has encouraged hydroelectric power development, first by government enterprises and later by pri-

vate companies—though not without serious and uncompensated impacts on other water users, particularly irrigators.

The Water Code has been much less effective in achieving other original objectives, particularly those having to do with the smooth operation of water markets and market incentives. This observation is now widely accepted by Chilean water experts, although not by the law's strongest political supporters. Note that some of these results represent a lack of success rather than failure; others are more clearly negative.

- Market incentives to promote more efficient water use, particularly within the agricultural sector, have not worked as expected. Irrigation efficiency remains low nationwide, and in the few areas where it has increased, the change reflects factors other than the water market—namely, investment to improve crop yields or reduce costs of labor and canal maintenance. Investment in these areas has been encouraged by the legal security of property rights, but not by market incentives to sell unused or surplus water rights; such rights are rarely sold.
- Continued government subsidies have been required for the construction and maintenance of irrigation works at small, medium, and large scales.
- Examples of significant market activity, as indicated by the amount of water resources reallocated or by the frequency of water rights transactions, remain limited to a few areas of the desert north and the metropolitan area of Santiago.
- The definitions of water rights remain vague or incomplete; the legal and technical details are inadequate and confusing in most of the country.
- The idea that water market forces would benefit peasants and poor farmers by improving their access to or ownership of water supplies has generally failed. If anything, the water market seems to have harmed many of these farmers more than it has helped them, although there is not enough evidence to make strong generalizations.
- Reliance on private bargaining to coordinate different water uses and resolve river basin conflicts, particularly between consumptive and nonconsumptive water rights, has failed. Neither DGA nor the courts have adequately or reliably redressed the problem.
- In the hydroelectric sector, nonconsumptive water rights have been subject to problems of speculation, concentrated ownership, and private monopoly power.

The mixed performance of Chilean water markets is due to the variety of factors that shape these markets' wider social, institutional, and geographic contexts. Institutional arrangements—the rules of the game—have been among the most important of these factors.[2] The Water Code's laissez faire definition of property rights has obviously had a strong impact on the specific economic incentives and disincentives that are faced by water users and water rights owners.

Comparison with Issues Ignored 20 Years Ago But Critical Today

The most negative results of the Water Code have involved issues that were of little concern in Chile in 1981 but have emerged as ever more critical since the early 1990s. These are the economic, environmental, and social problems that are at the heart of contemporary international debates about integrated water resources management and water governance:

- management of river basins, coordination of multiple water uses, and conjunctive management of surface water and groundwater;
- resolution of water conflicts through either judicial or nonjudicial processes;
- internalization of both economic and environmental externalities;
- clarification, enforcement, and monitoring of the relationships among different property rights and duties, such as consumptive and nonconsumptive water rights;
- environmental and ecosystem protection, including the maintenance of in-stream flows for environmental purposes; and
- public assistance to poor farmers to improve social equity in matters of water rights and water markets.

I described some examples of such problems in Chapter 4. Under the current Chilean institutional framework, these issues have generally been addressed in an ad hoc or ineffective manner and in some cases have not been addressed at all. Many of the flaws in the existing framework have been widely recognized by Chilean water experts, regardless of their political viewpoints. In some areas the empirical research is still missing or insufficient to draw definitive conclusions, although the available evidence indicates at least that there have been problems.[3]

Research on the Chilean Water Code has had strengths and weaknesses. The relatively single-minded focus on water markets and water rights trading means that we understand how they have worked much better than before, and this is an important step forward. This focus has also had significant blind spots, however. Our improved understanding of markets has come at the cost of other issues of water management and institutional performance that are at least as important. There is much more to water resources management than simply the allocation of resources—whatever the allocation mechanism may be.

Because the Water Code did not address the broader economic, environmental, and social problems that are important today, it may be unfair to criticize the code for its failure to solve them. But that is not the point here. Rather, the current legal and institutional framework—which is determined by the Constitution as well as by the Water Code—has shown itself incapable of handling these unforeseen problems. The current framework, as we have seen, is characterized by a combination of elements that reinforce each other to maintain the status quo: strong and broadly defined private economic rights; tightly restricted government regulatory authority; and a powerful but erratic judiciary that is untrained in public policy analysis, reluctant to intervene in issues with political implications, and committed to a narrow and formalistic conception of law. The problems of water management will only get worse as the demands and competition for water increase, putting ever more pressure on the existing institutional framework.

In view of the mixed results and the problems identified, what are the prospects for improving and modifying the current framework in Chile? Some kind of significant reform is essential, but the political and constitutional barriers to such reform are high. The Water Code's legal and institutional framework, shaped by its original focus on agriculture and rooted in the political and ideological context of the late 1970s, is locked in constitutionally.

POLITICAL DEBATE ABOUT REFORMING THE WATER CODE

By early 2004 the defenders of the Water Code in Chile had successfully resisted more than 10 years of government efforts at legislative reform.[4] Ever since Chile's return to democratic government in 1990, there has been continuing domestic political debate about how to modify or improve the water law. The core issue has been the definition of water rights: how to change the legal rules and economic incentives that affect private water use. Other issues of water policy, such as river basin management and in-stream flow protection, have been included in the proposals but have been overshadowed and eventually displaced by the heated controversy over property rights.

The Chilean government has won some tactical victories over its political opponents and has managed to keep the reform process alive, if not always moving forward. The bottom line, however, is that the Water Code was still intact in early 2004 and the prospects of change in the near future remain uncertain. And in any event, the terms of debate and the scope of the proposed reforms have narrowed considerably; even if the reforms are approved soon, the impacts on water management will be limited.

How can we explain this stalemate and lack of progress after so much expenditure of time and energy?[5] The government's agenda for reform has been undermined by two errors of political strategy. The first miscalculation was evident at the time: the design and presentation of the first round of proposed reforms in the early 1990s. In the context of contemporary Chilean politics and constitutional arrangements, those proposals were too aggressive in their criticism of the current model, and at the same time too vague in their legal details, to succeed. The biggest problems were the five-year use-it-or-lose-it rule for water rights and the proposal for new river basin organizations.

The government, led by DGA, apparently believed that political consensus about the need to reform the Water Code's basic principles made it unnecessary to woo the support of the opposition. When DGA put its blunt and rather poorly considered suggestions on the table, with the expectation of negotiating some of the details, many water users as well as ideological opponents resisted. The resulting polarization has had a lasting impact: it not only doomed the first round of reforms but also raised opponents' suspicions in a way that damaged the prospects for future cooperation.

The government's second miscalculation, in the mid to late 1990s, is more evident in hindsight: it seriously underestimated the ideological dogmatism and political intransigence of the opposition. DGA had learned from the failure of the first round of proposed reforms and made its second set of proposals more limited and pragmatic. In addition, the agency's overall diagnosis and analysis of water management issues, as described in the documents outlining the new National Water Policy, were generally balanced, comprehensive, and carefully considered. Agency staff showed a clear awareness and understanding of contemporary international debates about water policy and management. They refined the government's position on water markets to recognize the benefits of such markets as long as they were more adequately regulated. Unfortunately, the government overestimated the opposition's willingness to negotiate and spent its political capital trying to achieve a relatively minor initial step in the reform process—establishing fees for nonuse of water rights.

Since the merits of those fees are debatable, it is doubly unfortunate that they have dominated public debate about water law ever since. Fees for nonuse have some drawbacks as an instrument of public policy, as discussed in Chapter 3, but in my view, they would have been acceptable if they could have been easily approved. Instead, so much attention was focused on the fees for nonuse that more important reforms were pushed aside.

How did the government end up putting so much effort into so minor and flawed a reform measure? Why were the reformers unable or unwilling to reassess their approach along the way? It seems that at many points in the long and drawn-out political debate, leaders in the government thought that Congress was about to approve the fees for nonuse and then they could move on to the issues that really mattered. A related factor was the momentum and political capital that the government had invested in its chosen strategy: the more time passed, the more difficult a change of course. And finally, the government was motivated by regulatory problems in the electricity sector, where issues of private monopoly power and concentrated ownership of nonconsumptive water rights affected the debate about the Water Code. These electricity issues, in fact, often had more public and political visibility than water issues.

In retrospect, the government should probably have abandoned the proposed fees for nonuse in 1997 or 1998, when the reform's opponents had made clear that they would fight the proposal tooth and nail (despite their hints to the contrary a few years earlier). At that point the government could have called the opposition's bluff by agreeing to the alternative of water rights taxes. Even if such taxes could not really be implemented, for all the practical and political reasons discussed in Chapter 3, this strategy would have taken away the antireformers' easiest political target and most effective policy argument and perhaps forced them into a genuine dialogue about the government's proposals.

From the standpoint of many of the reform's opponents, of course, the lack of progress over the past decade has been a political victory. In this respect it is important to distinguish between two tendencies within the opposition. On the one hand, some of the arguments against the proposed reforms have been reasonable and well founded and have pointed to weaknesses or inconsistencies in the government's arguments, as summarized at the end of Chapter 3.[6] Skepticism about the government's true position on the role of markets in water management is understandable, since this position has sometimes been ambivalent. Property owners in Chile, after all, have sound and fairly recent historical reasons to distrust government agencies with broad discretionary powers (although the current Constitution and institutional framework provide powerful controls to prevent that history from being repeated).

On the other hand, many opponents of Water Code reform have been highly ideological, politically intransigent, and unwilling to compromise. Their attitude has been evident throughout the 1990s and if anything it has gotten stronger over time, perhaps because it has been politically effective. Many of the antireformers' policy arguments have been political tactics rather than constructive criticisms or counterproposals. Despite their rhetoric about the need for reform and improved economic instruments, at the end of the day these opponents have not been willing to agree to any significant changes. Criticism of the government's allegedly antimarket position has been greatly exaggerated for political effect and reflects the highly ideological nature of political debate in Chile, in which anyone who is not a free-market neoliberal can be accused of being antimarket. The best example of the antireformers' combination of political skill and insincerity is their argument in favor of water rights taxes, a reform measure that they would be very unlikely to ever support in practice.

Thus the Water Code reforms have failed because the 1980 Constitution and its associated institutional arrangements are so strongly weighted in favor of right-wing political interests, laissez faire economics, and vested property rights. No meaningful legal reform is possible without the support of those same political and economic interests. Representatives of private companies and private sector interest groups will sometimes admit (off the record) that their opposition to all taxes or fees on water rights cannot be justified by economic theory, but they nonetheless defend the status quo because it has brought them great material benefits. The lack of participation in the political debate by citizens and nongovernmental organizations (NGOs) representing environmental interests, peasant farmers, or indigenous peoples has further weakened the possibility of reform. In all these respects, the attempts to reform the Water Code have been a microcosm of the Chilean transition to democracy in general.

Although the current stalemate is in some sense a victory for the Water Code's defenders, it leaves pending the broader issues of integrated and sustainable water resources management in Chile. The ability of the Water Code's institutional framework to resist reform thus far may make it politically impossible to implement more ambitious reforms in the areas of environmental flows, management of river basins, and resolution of water conflicts, at least in the foreseeable future. Those reforms would involve a stronger challenge to the current model.

Moreover, the obstacles to reform are even higher now because the balance of political forces in Chile has been shifting to the right and the national economy has weakened since the late 1990s. National congressional elections in December 2001 increased the representation of right-wing political parties in both the Chamber of Deputies and the Senate. The current political and economic crises in Argentina and other Latin American countries have made Chilean politicians—those in the governing Concertación as well as those in the opposition—more nervous about attracting foreign investment and encouraging renewed economic growth.

In the present context, therefore, people committed to improving Chilean water policy and management will have to create greater capacities for persuasion, dialogue, pragmatism, and long-term vision. Part of the political strategy should be to take advantage of the variety of perspectives and interests within the Chilean right—particularly the differences between neoliberals and more traditional conservatives, and between sectoral interests in water use (agriculture, mining, electricity, urban development) in different parts of the country.

What should be the next steps in the Chilean reform process? In my view, the points and priorities should be the following:

- The government should be willing to drop the proposed fees for nonuse. They are not a big enough prize to keep fighting for, and they are not worth the sacrifice of more important water policy objectives if that is the political price to be paid. Politically, it is probably too late for the government to withdraw this proposal unilaterally, unless it is part of a renewed effort to negotiate with the opposition about water rights taxes.
- The government should join the opposition in declaring water rights taxes the medium-term goal, while recognizing the practical difficulties of implementing such a system in the near future. Hence the declaration of policy should also describe a five-year or ten-year plan for implementation, including institutional strengthening and capacity building. This

plan would build on recent government efforts to regularize water rights titles and better define water rights throughout the country but would include a more ambitious and systematic program in the future. The plan should also strengthen existing government programs in favor of small-scale and peasant irrigation systems. The revenues from the new water rights taxes should be earmarked for water resources management and protection.

- Both government and opposition should recognize that the legal requirement to *use* water rights is not in itself the essential point, and the lack of such a requirement in the Water Code is not the critical problem. Instead, the critical problem is that property rights to water are defined as strictly private commodities in such broad and unconditional terms that there is no effective way to assert or defend public rights and interests—whether these public interests are economic, social, or environmental. This is what leads me to favor water rights taxes: not because I think they are the best policy in theory, but because I think they are the most feasible way to assert public interests in the current Chilean context.
- Legislation should be drafted to clarify the rules governing the exercise of nonconsumptive versus other water rights in managing river basins, dams, and reservoirs. This legislation should include explicit clarification of the authority of the largest water users' organizations (the vigilance committees) and of DGA to enforce these rules.
- The continuing attempts to design new, comprehensive, public-private river basin organizations should be abandoned. Instead, the focus should be on drafting and negotiating legal agreements or compacts for specific basins. These compacts might include some form of new basin agency to implement the agreement, but only after the hard decisions about institutional design have been made.
- Protection of in-stream flows for environmental purposes should be considered in the context of the two preceding items about river basin management. I assume here that the Chilean government will maintain its present division of regulatory authority between DGA (with jurisdiction primarily over water quantity and water resources management) and the National Environment Commission (which handles environmental protection and some aspects of water quality). Even so, in-stream flow protection must be addressed in the context of managing river basins.
- Environmental and social NGOs should be encouraged to participate more actively and in a more informed manner in public policy debates

about water law and water resources management. The presence of such NGOs is essential to articulate various public interests independent of government agencies and officials. This will require additional training of NGO staff and other forms of financial support.

- The Chilean government and Chilean universities should undertake a sustained program of applied research and policy analysis about alternative institutional arrangements for resolving water resource conflicts, including environmental issues.[7] The program should be interdisciplinary and comparative and include both existing and potential institutional alternatives. The alternative arrangements to be studied should include, at a minimum: (1) increased administrative discretion and regulatory capacity, together with increased review and oversight by some sort of water commission or special courts; (2) training and reform within the ordinary judicial system; and (3) some form of arbitration or alternative dispute resolution. This research program will presumably require some financial support from, and other collaboration with, international organizations.

Achieving political consensus on reforms will obviously be very difficult, and perhaps impossible, in the foreseeable future. The country's current political and institutional context presents major obstacles, and how people in Chile will overcome them and address the problems remains to be seen.

This situation, however, with all its difficulties and constraints, should be widely publicized in other countries that are interested in the Chilean water rights system. The Chilean experience underlines the political nature of economic instruments—both in the initial decision to adopt them and in any subsequent attempts to modify them.

INTERNATIONAL WATER POLICY: LESSONS FOR REFORMS

Chile's experience shows the problems that can flow from implementing a free-market water law. Its narrow economic approach has led to policies and institutional arrangements that cannot meet the challenges of integrated and sustainable water resources management. To avoid such outcomes, international efforts to reform water policies must foster a broader and more interdisciplinary approach to water economics, with more legal, institutional, and political analyses of markets and economic instruments.

My hope is that this analysis of the Chilean experience will help raise the level of international debate about IWRM, particularly about its economic and institutional aspects.

The Chilean model has had two main economic benefits: first, the legal security of private property rights has encouraged private investment in water use, for both agricultural and nonagricultural uses; and second, the freedom to buy and sell water rights has led to the reallocation of water resources to higher-value uses in certain areas and under certain circumstances. These are important benefits, even though market incentives themselves have been only partly functional in practice, and they are the kind of results that promarket policies hope to deliver. Many other countries would improve their water use and management if they could achieve similar results.

However, those benefits are directly linked to a legal, regulatory, and constitutional framework that has proven not only rigid and resistant to change but also incapable of handling the complex problems of river basin management, water conflicts, and environmental protection. These more complex problems, of course, are precisely the fundamental challenges of integrated water resources management. In addition, peasants and poor farmers have for the most part not received the economic benefits, which indicates that social equity is another weak point of the current framework.

The strengths of the Chilean model, in other words, are also its weaknesses: the same legal and institutional features that have led to the model's success in some areas have effectively guaranteed its failure in others. The ability of the institutional framework to resist reform is the strength of rigidity. Its regulatory limitations, incomplete price signals, and lack of balance have been built to last. Even if we put the most positive spin on the model's economic benefits, for the nation as a whole these flaws are a high price to pay, particularly over the long term.

This analysis points to the inaccurate manner in which the Chilean model has often been described in international water policy circles, particularly by the economists who have been the model's strongest supporters. The Chilean experience is cited as an example of successful free-market reforms, and although supporters may recognize some problems in the model, they present them as secondary issues that do not affect the overall positive assessment. Hence the absence of effective institutions for managing river basins or resolving water conflicts is mentioned as an afterthought, or played down as a separate issue that can be addressed later, or seen as acceptable in light of the model's presumed advantages. According to this viewpoint, the problems have been identified and the Chilean government is in the process of solving them. In this way the spotlight remains on water markets and water rights trading, supposedly the model's strong points. The message to other countries is that they can adopt the Chilean model of water markets without also adopting its deeper institutional weaknesses in other areas of water resources management.[8]

Such assessments are mistaken and misleading and based on insufficient knowledge about Chilean politics and institutions. The flaws in the Chilean model are structural: they are integral parts of the same legal and institutional arrangements that underlie the water market. These flaws are not separable from the rest of the model. On the contrary, they are the necessary institutional consequences of the free-market reforms of property rights and government regulation. The aspects of the model that privatize water rights so unconditionally and define them as freely tradable commodities are inextricably connected to the aspects that weaken and restrict the regulatory framework. This is not a theoretical matter: in Chile these structural connections have been demonstrated in practice over the past 20 years, both by the mixed empirical results of the Water Code (see Chapter 4), and by the long and fruitless process of attempted Water Code reform (see Chapter 3).

Another way of putting this argument is to return to the image of IWRM and sustainable development as a tripod whose three legs are economic efficiency and growth, social equity, and environmental sustainability. The Chilean model of water management has one strong economic leg and two weak social and environmental legs, making it unbalanced overall. The social and environmental legs cannot be strengthened without weakening the economic leg in ways that—at least in Chile—are politically and constitutionally difficult. Moreover, even the economic leg is weaker than it appears because the ineffective mechanisms for resolving conflicts and internalizing externalities also reduce economic efficiency and growth, especially over the long term. Because the Chilean approach to managing water as an economic good puts all the emphasis on water as a *private* good and tradable commodity, it is very difficult to recognize or enforce the other aspects of water as a *public* good.[9]

For people in other countries and in international organizations who are interested in water law and policy reforms, this offers sobering lessons. The most obvious warning is that other countries should not copy or closely follow the Chilean water law model, at least not without a thorough understanding of the model's weaknesses as well as its strengths. Both the flaws and the rigidity of the Chilean institutional framework should be made clear to other countries interested in the Chilean approach. The problems for IWRM should not be presented as secondary, separate from the water market, or readily solved. The failure of the World Bank and other boosters to make this clear in their advocacy of the Chilean model has been highly irresponsible.

Many of the details of the Chilean model are unique to Chile and have been shaped by the local political and economic context, particularly by the Chilean Constitution. It is also probable that this document has an unusual weight and importance in Chile that is matched by few constitutions in other countries, since the Chilean Constitution dictates the basic rules for economic institutions as well as for the political system. It might seem possible to create an "improved" version of the Chilean Water Code by keeping the model's better features and avoiding its more serious flaws. If Chile went too far in the free-market direction and is now locked in place by its history, politics, and constitution, that is Chile's problem—it does not prevent other countries from learning from Chile's mistakes.

From an institutional perspective, however, the prospects for such an improved version in other countries are slim. Regardless of the specifics of the Chilean case, any country that tries to follow the laissez faire economic approach of Chilean water law will necessarily confront similar institutional and political problems. How is it possible to create a legal and institutional framework that provides such strong guarantees for private property and economic freedom, and such wide scope for free trading of water rights and private decisionmaking about water use, without also severely restricting government regulation and legislative reform? If a country does not want to grant the judiciary such broad powers to review the actions of government agencies, how else can those agencies be prevented from interfering in water markets and property rights? If private economic rights are so strong and public regulation is so weak, through what institutional mechanisms other than the courts can conflicts be resolved effectively? How much room can there be for environmental protection in such a framework, and how can the level of that protection increase over time?

If other countries want to follow the Chilean approach to water economics, they will have to adopt a legal and institutional framework that is functionally equivalent to Chile's. If instead a country chooses a stronger regulatory framework or places more conditions on private rights, that country is, by definition, no longer following the Chilean economic approach. Hence one of the deeper lessons of the Chilean water model is to show how different economic perspectives have different consequences for institutional design. The Chilean experience shows the lasting problems that result when a narrow economic perspective is combined with the political power to design legal institutions in its own image.

In international debates about integrated water resources management, the principle that water should be recognized as an economic good should not be thought of as a separate or independent component of legal and policy reforms. In particular, countries and governments should not make the mistake of thinking that they can implement reforms in two steps, by first adopting a free-market approach to water economics as a straightforward initial step, and then turning their attention to the remaining problems of IWRM and water governance. At that later point their hands will already be tied by a definition of property rights that has major political and institutional implications. On the contrary, reformers should put greater and earlier efforts into mechanisms for resolving conflicts and reflect carefully on how to define and enforce property rights to something as complicated as water.

We do not want to throw the baby out with the bathwater. I am arguing against "free" markets and narrow economics, not against all use of market-based instruments and analyses in water management. The goal of a broader and more interdisciplinary approach to water economics is to build on conventional neoclassical principles, not to reject them. We can get the most benefits from markets by recognizing their limits and not asking them to handle problems beyond their scope. That is why, in my view, we need to revive older approaches to institutional economics—approaches that are rooted in qualitative and historical analyses and draw on politics, law, and other social sciences in order to understand "economic" issues.[10]

The combination of such institutional economics with the study of law in social context, or "law in action," is what I refer to as comparative law and economics. This approach, summarized in Chapter 1, is inherently qualitative and interdisciplinary simply because it is comparative, and I have tried to apply it throughout this book. It is especially important in developing countries, where institutional and social contexts are significantly different from the developed countries where most economic theory has originated and continues to be shaped. Comparative law and economics can be applied to far more than water policy, of course—it is an approach that would improve discussion about many other critical issues of international development. One example is the recent debate about the "second generation" of institutional reforms needed to correct the problems caused, or not solved, by the first generation of neoliberal economic reforms promoted by the Washington Consensus.[11]

The current global water crisis is driven by scarcity and conflict, two problems that are ever more tightly bound together. Although economic principles can be powerful tools for dealing with water *scarcity*, legal and political institutions are the key to resolving water *conflicts*. The Chilean experience confirms the need for a more interdisciplinary perspective on law and economics in the design of policy reforms.

NOTES

Introduction. The Chilean Water Model Comes of Age

1. See, for example, Birdsall et al. 1998; Graham and Naím 1998; Inter-American Development Bank 1999. For examples of the World Bank's attention to judicial reform in this context, see Dakolias 1996 and Rowat et al. 1995.

Chapter 1. The International Context

1. For example, the World Water Council refers to "a chronic, pernicious crisis in the world's water resources" (Cosgrove and Rijsberman 2000, Preface); the Global Water Partnership refers to "a looming water crisis" (Global Water Partnership 2000a, 11); and the title of Peter Gleick's edited book, *Water in Crisis*, speaks for itself (Gleick 1997).
2. Global Water Partnership 2000b, 22. As discussed below, the partnership is an international organization dedicated to promoting integrated water resources management around the world.
3. For representative examples of contemporary international debate about water resources management, see Dourojeanni 1994; Food and Agriculture Organization et al. 1995; García 1998; Gleick 1998; Lord and Israel 1996; World Bank 1993; and the citations in the following two notes. These publications vary in some of their emphases and recommendations, but they all share the same general features of seeking to combine social, economic, environmental, and other aspects of water use and management.

4. Dublin Statement, International Conference on Water and Environment 1992, 4.
5. See Global Water Partnership 1998, 2000a, 2000b; Solanes 1998; Solanes and González 1999. The partnership's headquarters are in Stockholm, and it has regional offices throughout the world.
6. Global Water Partnership 2000a, 23.
7. Both definitions are quoted in Rogers and Hall 2002, 4 (a Global Water Partnership publication).
8. Dublin Statement, International Conference on Water and Environment 1992, 4.
9. For example, the World Water Council's "vision," prepared for the forum at The Hague, spoke of the vital need for "pricing of water services at full cost" (Cosgrove and Rijsberman 2000, Executive Summary). The final report for that forum listed four major issues about which there was significant disagreement: "privatization, charging the full-cost price for water services, rights to access, and participation" (World Commission on Water for the 21st Century 2000, 16). For examples of critiques of privatization by nongovernmental organizations, see Barlow 2000; Gleick et al. 2002; Public Services International 2000.
10. It should be clear that I am not referring to the "new" institutional economics or to the rational choice approach to political economy, both of which are essentially applications of neoclassical economic theory and methods to institutions and politics. Similarly, the field commonly known as "law and economics," a term that implies a combination of the two disciplines, would more accurately be called neoclassical economic analysis of law. Although these approaches may offer useful insights, they nevertheless reflect narrow rather than broad perspectives on economics. See also the following note.
11. This book is not the place to review in detail the vast literature about institutional economics and law and economics. Some examples will be discussed later in this and subsequent chapters. A range of useful references, from various perspectives, would include Bardhan 1989; Barzel 1989; Bromley 1989; Coase 1988; Cole and Grossman 2002; Commons 1924, 1934; Field 1981; Hodgson 1988; Libecap 1989; Mercuro and Medema 1997; North 1981; Rutherford 2001; Williamson 1985.
12. World Bank 1993. For a careful and comprehensive study of the World Bank on the occasion of the 50th anniversary of its founding—beyond the specific issues of water resources—see Kapur et al. 1997.
13. Rosegrant and Binswanger 1994, 1619, 1623.
14. Easter et al. 1998, 8, 280. Although this book is not a World Bank publication, all three of its editors are economists who are either employed at or closely associated with the Bank.
15. For example, some of these mitigating strategies are as follows: "Require review and approval of transactions by public agency.... Establish a fund to compensate third-parties damaged in trading, financed by levies on water transactions.... Revise water rights downward.... Open litigation to non-holders of water rights.... Tax or ban on trading from upstream to downstream users.... Set minimum instream flows to maintain aquatic ecosystems.... Tax unused water rights.... Aid small water rights holders with free legal protection and information.... Define two types of rights with one senior to the other." Ibid., 278–79.
16. Ibid., 282.
17. For excellent examples of comparative sociolegal analysis of water rights and conflict resolution, see Benda-Beckmann et al. 1997; Bruns and Meinzen-Dick 2000. On the law and society tradition in general, see the journals *Law and Society Review*, published by the Law and Society Association, and *Law and History Review*, published for the American Society for Legal History.

18. In the papers discussed here, Briscoe cautions that his opinions are not to be taken as expressing the official views of the World Bank. Nonetheless, we will see later that other Bank-associated publications present very similar, if not identical, positions.
19. Briscoe 1996a, 1997.
20. Briscoe 1996a, 9–10.
21. Ibid., 2, 17.
22. Briscoe 1997, 6.
23. Briscoe 1996a, 22.
24. Rogers et al. 1998. This paper was written in 1996 and published by the GWP two years later. On the GWP, see Global Water Partnership 1998, 2000a, 2000b; Solanes 1998; Solanes and González 1999.
25. Rogers et al. 1998, 6–10. See especially their Figure 1 on page 7.
26. Ibid., 10–14. See their Figure 2 on page 13.
27. Ibid., 14.
28. Ibid., 14, 17.
29. Perry et al. 1997, 1.
30. Ibid., 16–17. In terms of specific policy implications, in the area of international irrigation development their analysis leads them to favor government subsidies rather than market prices, in the interest of improving social equity.
31. McNeill 1998, 253–54, 256. Like Briscoe and Rogers et al. (as cited above), McNeill describes the value of water as the "money costs" of providing water to particular users, including operation and maintenance (what Briscoe calls use costs and what Rogers et al. call full supply costs), plus the opportunity cost—that is, water's value in alternative uses. He points out that opportunity costs are often very hard to measure in monetary terms.
32. Ibid., 258–60.
33. Brown 1997, 3–4. Emphasis in the original. Brown has worked on issues of competing perspectives on the value of water resources for many years; see, for example, Brown et al. 1982; Brown and Ingram 1987.
34. Bromley 1982. The four theoretical concepts examined in this paper are "Pareto optimality, the notion of Pareto-irrelevant externalities, the intertemporal allocation decisions of private resource owners, and the scientific basis for decisions about particular institutional arrangements" (834). In subsequent books Bromley develops his arguments at much greater length; see Bromley 1989, 1991.
35. As he explains elsewhere (Bromley 1989, 4), "Efficiency calculations rest upon the current structure of institutional arrangements that determine what is a cost—and for whom .... To identify the *efficient* policy choice against which others are to be compared is to load the debate. There is *no single efficient policy choice* but rather an efficient policy choice for every possible presumed institutional setup. To select one efficient outcome is also to select one particular structure of institutional arrangements and its corresponding distribution of income. What matters is not efficiency, but *efficiency for whom?*" (Italics in original.)
36. Bromley 1982, 842–43. His emphasis on the importance of law for determining value builds on the work of John Commons, a leading institutional economist of the early 20th century; see Commons 1924, 1934. For other U.S. applications of institutional economics to water resources, including discussions of the contrasts with neoclassical economics, see Ciriacy-Wantrup 1967; Livingston 1993a, 1993b; Miller et al. 1996, 1997; Wandschneider 1986.
37. Aguilera 1998. For the similarities with the United States, see Note 54 below and accompanying text.

38. Ibid., 19.

39. In his latest book, which is part of a series of publications aimed at promoting a "new water culture" in Spain, Aguilera takes an institutional economic approach to studying water markets in the Canary Islands of Spain; see Aguilera and Sánchez 2002. Aguilera's innovative and sophisticated work is representative of a recent wave of important publications on Spanish water issues by institutional and ecological economists; see, for example, Aguilera et al. 2000; Arrojo 1995; Arrojo and Naredo 1997; and other papers published in the same conference proceedings as Aguilera 1998. See also the following note.

40. Ecological economics is a diverse and eclectic field whose basic critique is that the neoclassical framework ignores fundamental aspects of the physical and ecological sciences, particularly the flows of energy and matter through ecosystems. Hence conventional economic analyses tend to treat economic systems as if they were largely isolated from the physical environment and its dynamics and constraints. In this book I do not discuss or apply the specific framework of ecological economics, although my own background in geography and geology means that I share the basic perspective. For more detailed discussions of ecological economics, see, for example, Costanza et al. 1996; Daly and Townsend 1993; Prugh et al. 1999. For an application to water law and policy in particular, see Young 1997.

41. Because the Chilean Water Code was written more than a decade before the international water conferences and policy debates summarized earlier, the Chilean law itself was evidently shaped by factors other than the Dublin Principles, as we will see in Chapter 2. Nonetheless, all the international discussion of the Chilean model has taken place since 1990, in the context of those recent international debates. Similarly, within Chile most discussion of water law and policy since 1990 has referred explicitly to the international debates.

42. See Rosegrant and Binswanger 1994; World Bank 1994; Hearne and Easter 1995; Ríos and Quiroz 1995; Simpson and Ringskog 1997; Easter et al. 1998. Although the 1998 book is not a World Bank publication, all three of its editors are economists who are either employed at or closely associated with the Bank.

43. See Dourojeanni and Jouravlev 1999; Solanes 1996. See Food and Agriculture Organization 1999 for a somewhat less critical assessment.

44. See Trawick 2003.

45. All the countries named are examples about which I have some personal knowledge, based on my work as an international consultant, participation in international meetings and conferences, and personal communication with water experts in the countries involved. There are almost certainly other examples.

46. Briscoe 1996a, 21–22; Briscoe 1997, passim; Briscoe et al. 1998. The last paper, although coauthored with two senior Chilean government officials, draws heavily on Briscoe 1996b.

47. Briscoe 1996a, 21 (emphases added). The one example is the Limarí River, which I discuss in Chapter 4.

48. Ibid., 20. This distinction is revealing about the limitations of the Chilean model for integrated water resources management, since a basic principle of IWRM is that water quantity and water quality are intimately related and should not be managed by different institutional frameworks. Briscoe does not point this out, however.

49. Briscoe et al. 1998, 9 (emphasis added).

50. Rogers and Hall 2002, 25–26 (emphasis added). Their source for this assessment is Briscoe et al. 1998, as indicated in Rogers 2002, 47–50, 54.

51. See, for example, Dubash 2002; Meinzen-Dick 1996; Shah 1993.

52. See Challen 2001; Clark 1999; Young 1997; and several papers in *Water International* 24(4) 1999.

53. See Aguilera 1998; Aguilera et al. 2000; Aguilera and Sánchez 2002; Arrojo 1995; Arrojo and Naredo 1997; Del Moral and Sauri 1999; Embid 1996; and the special issue of *Water International* 28(3) 2003.
54. The literature about U.S. water markets, law, and policy is vast. For overviews and representative examples, see Bates et al. 1993; Bauer 1996; Brown et al. 1982; Fort 1999; Frederick 1986; Getches 1996; Hildreth 1999; Howe 2000; Saliba and Bush 1987; Shupe et al. 1989; Tarlock 1991; Willey 1992.
55. E.g. Wescoat 2002.

Chapter 2. The Free-Market Model

1. In this book I do not discuss the long history of Chilean water law before 1951, when the country's first Water Code was passed. For more information about the Spanish colonial era and the 19th and early 20th centuries, see Bauer 1998b, 36–38, and citations therein.
2. The following summary is taken from Bauer 1998b, 33–36, which has additional detail and citations.
3. The Water Code applies only to terrestrial waters, not to oceans or coastal waters. The Civil Code also defined certain categories of waters to be privately owned—namely, small streams and water bodies contained within a single landed property, and more importantly, waters flowing in "artificial" channels (canals). The latter rule was intended to provide incentives for private irrigation development by recognizing rights to waters once they had been diverted from rivers. See Bauer 1998b, 47, Notes 16, 17.
4. The 1981 code also recognizes all water rights granted or acquired under previous legislation, and its Transitory Articles establish legal procedures to "regularize" the confusing and uncertain status of many preexisting water rights titles. In theory, all water rights must now be measured in terms of volume per unit of time, such as liters per second, but in practice many older rights are expressed as proportional shares of available flows or by other measures.
5. Solanes 1996.
6. The legal distinction between public ownership of water and private ownership of rights to use that water has long been controversial and confusing in Chile. Many Chilean water lawyers, even if they favor markets, have considered the current Water Code legally incoherent because it effectively privatizes a resource that it simultaneously defines to be inalienably public. Other lawyers and most economists defend the code's approach as the only practical way to adapt the peculiar physical characteristics of a fluid, moving resource to the logic and requirements of a market. See Bauer 1998b, 46, Notes 6, 7.
7. The Water Code does give DGA authority to order redistribution of water rights during officially declared drought emergencies, but even so the government must compensate water users who are damaged as a result (Art. 314). To my knowledge this has never happened.
8. This theorem was named after Ronald Coase, Nobel Laureate at the University of Chicago, and is one of the pillars of the field of law and economics. See the works cited in Chapter 1, Note 11.
9. For a more detailed analysis of nonconsumptive rights and the associated legal and institutional problems, see Bauer 1998a and Bauer 1998b, 79–118.
10. Some of the Constitution's most authoritarian political features were removed or modified in 1989 in a package of reforms negotiated between the military government, the conservative political party Renovación Nacional, and the incoming coalition government of the Concertación. Nonetheless, the basic framework remained intact, and in particular

the economic and regulatory aspects were untouched. See Bauer 1998b, 11–12, and accompanying footnotes.

11. For detailed accounts of Chilean political history before and during the military government, see Angell 1993; Constable and Valenzuela 1991; Drake and Jaksic 1991; Loveman 1988; Valenzuela 1989, 1991. A very brief summary of recent Chilean political and economic history can be found in Bauer 1998b, 3–5.
12. For further discussion of the nature and significance of the 1980 Constitution and the expanded powers of the judiciary, see Bauer 1998b, 11–31, and Bauer 1998c and the citations therein. For detailed accounts of the internal political, legislative, and policymaking processes during the military government, see Barros 2002; Cavallo et al. 1989; Fontaine 1988; Valenzuela 1991.
13. For the general political context, see the works cited in the previous two footnotes. This chapter's section on the 1981 Water Code's "legislative history" is taken from Bauer 1998b, 40–45, which has more complete citations.
14. Law 9,909, published in the *Diario Oficial* on 28 May 1951. For more detail and citations, see Bauer 1998b, 38.
15. See, for example, Sax et al. 1991.
16. Indeed, the current legal term for water rights, "rights of advantageous use," with its implicit conditionality—what if the use is not "advantageous"?—dates from the 1951 code.
17. For more detail and citations, see Bauer 1998b, 39–40.
18. A great deal has been written about the Chilean Agrarian Reform. Good overviews are Garrido et al. 1990; Jarvis 1985, 1988.
19. The Agrarian Reform Law was Law 16,640, passed in 1967. The new Water Code took effect at the same time but was published separately in 1969 as Decree with Force of Law 162. I will refer to it as the 1967 Water Code to emphasize its legal status and links with the Agrarian Reform. For further references about this code, see Bauer 1998b, 47, Note 24.
20. For some legal fine points about the nominal compensation that was offered, see Bauer 1998b, 48, Notes 25, 26.
21. For historical accounts of this period, see Angell 1993; Loveman 1988; Valenzuela 1989.
22. See Garrido et al. 1990; Jarvis 1985, 1988.
23. See Bauer 1998b, 11–25 and citations therein. The term in Spanish is *los Chicago Boys*.
24. For more on this commission, see Bauer 1998b, 12–19 and citations therein. The commission's sessions on water rights were later republished in *Revista de Derecho de Minas y Aguas* Vol. 1 (1990): 227–59.
25. Constitution (1980), Article 19, No. 24.
26. The 1976 Constitutional Acts were interim amendments of the 1925 Constitution, dictated by the *junta* while the new constitution was still in preparation. See Bauer 1998b, 26, Note 9. The *junta* itself had four members: the commanders of Chile's three armed forces (army, navy, and air force) and the national police force (Carabineros).
27. See *Acta* 280, 3 September 1976, *Actas de Sesiones de la Honorable Junta de Gobierno* (unpublished).
28. See Bauer 1998b, 11–25 and citations therein.
29. This Decree Law was published in the *Diario Oficial* on 23 April 1979. For additional references, see Bauer 1998b, 48, Note 36.
30. On the Chilean government's historical role in irrigation development, see Bauer 1998b, 42 and 46, Note 10.

31. For the economic arguments in favor of the new law, see citations in Bauer 1998b, 49, Notes 39 and 40.
32. Compare the proposed draft with the final version, in *Decretos Leyes Dictadas por la Honorable Junta de Gobierno: Transcripción y Antecedentes, Tomo* 167, *Folio* 1-356, in the Library of the National Congress.
33. See *Acta* 364, 7 February 1979, *Actas de Sesiones de la Honorable Junta de Gobierno* (unpublished).
34. Ibid.
35. Compare the proposed draft with the final version, in *Decretos Leyes Dictadas por la Honorable Junta de Gobierno: Transcripción y Antecedentes, Tomo* 167, *Folio* 1-356, in the Library of the National Congress.
36. Bauer 1998b, 49, Note 44.
37. "*Nuevo Código de Aguas impulsará la inversión,*" interview with Rule Bismarck, *El Mercurio,* 31 October 1981, C3.
38. The Water Code is Decree with Force of Law 1,122, the irrigation law is Decree with Force of Law 1,123, and both were published the same day in the *Diario Oficial.*
39. Law 18,450, which is still in effect. See Bauer 1998b, 76, Note 42, and accompanying text.
40. See Ellenberg 1980; Figueroa 1989; Instituto de Ingenieros 1993.

Chapter 3. Reforming the Reform?

1. On this period, see, for example, Angell 1993; Constable and Valenzuela 1991; Drake and Jaksic 1991.
2. "Lots of noise, but few nuts" is a popular Chilean phrase, roughly equivalent to Macbeth's "full of sound and fury, signifying nothing."
3. For more detailed discussion of the political and constitutional framework, see Bauer 1998b, 3–5, 11–25, and accompanying notes.
4. In addition to the documents cited, my analysis in this chapter is based on numerous interviews in Chile with government officials, water experts, and stakeholders from 1992 through 2003. Many of these interviews were at least partly off the record.
5. Manríquez 1993; interview with Gustavo Manríquez, 5 February 1992, Santiago.
6. Dirección General de Aguas 1991a, 2, and 1991b, 267.
7. See Chapter 2.
8. *Proyecto de Ley que Modifica el Código de Aguas, Boletín* 876-09, 1 (2 December 1992).
9. Ibid., 3–8.
10. See Chapter 2. As I mentioned previously, the use-it-or-lose-it rule is equivalent to the beneficial use doctrine in U.S. water law.
11. 1980 Constitution, Article 19, No. 24. See Chapter 2 for more detailed discussion of the constitutional status of water rights.
12. For a very similar theoretical argument made by an economist about water rights in the United States, see Barzel 1989, 89–91.
13. Important examples of opposition to the government's proposal are Donoso 1994; ENDESA 1993; Figueroa 1993a, 1993b; Instituto de Ingenieros 1993; Instituto Libertad y Desarrollo 1993a, 1993b; Sociedad Nacional de Agricultura 1993; "Retroceso en Régimen de Aguas," editorial in *El Mercurio,* 19 February 1993, A3; and other citations in Bauer 1998b, 77, Note 57.

14. *Proyecto de Ley que Modifica el Código de Aguas, Boletín* 876-09 (2 December 1992).
15. See, for example, ENDESA 1993; Instituto de Ingenieros 1993; Instituto Libertad y Desarrollo 1993a; Sociedad Nacional de Agricultura 1993.
16. For arguments in favor of water rights taxes in this first round of debate about Water Code reform, see Confederación de Canalistas de Chile 1993a; Figueroa 1993a, 1993b; Instituto Libertad y Desarrollo 1993b; Urquidi 1994; "Retroceso en Régimen de Aguas," editorial in *El Mercurio*, 19 February 1993, A3.
17. For arguments in favor of *patentes* for water rights in this round of the debate, see ENDESA 1993; Instituto de Ingenieros 1993; Sociedad Nacional de Agricultura 1993. On Chilean mining law, see Vergara 1992.
18. Dirección General de Aguas 1991b; Muñoz 1991. I will use *fee for nonuse* instead of *patente* because Chilean terminology is not always consistent: some people use *patente* to mean simply "fee," which can be a fee for either use or nonuse.
19. Gustavo Manríquez, Session 2, 10 March 1993, *Comisión Encargada del Régimen Jurídico de las Aguas*, Chamber of Deputies, 19.
20. Presidential Message 79-327 to the President of the Chamber of Deputies, *Formula Indicaciones al Proyecto de Ley que Modifica el Código de Aguas*, 30 September 1993.
21. Lagos 1994. Also interviews with DGA head Humberto Peña, 20 October, 7 December, and 18 December 1995.
22. Anguita 1995; Dirección de Riego, *Indicación Sustitutiva al Proyecto de Modificación del Código de Aguas*, memo to DGA, May 1995. Also interview with Pablo Anguita, 18 December 1995.
23. Law 19,300, *Ley de Bases del Medio Ambiente*.
24. For legal and political analysis of the 1994 law, the National Environment Commission, and the institutional framework in general, see Asenjo 1990; García 1999; Ruthenberg et al. 2001; Silva 1994, 1997.
25. Presidential Message 005-333 to President of Chamber of Deputies, *Formula Indicación al Proyecto de Modificación del Código de Aguas*, 4 July 1996.
26. Dirección General de Aguas 1999. Much of the same material can also be found in earlier papers and presentations by the head of DGA, Humberto Peña; see Peña 1997a, 1997b. I discuss these documents further in Chapter 4.
27. See, for example, Vergara 1997a for a report about how to design a national registry of water rights titles, and Donoso 2000 for an analysis about whether to adopt a system of charging fees for water use. Both Vergara and Donoso are leading Chilean experts on water law and economics, professors at the Catholic University, and supporters of the 1981 Water Code.
28. Presidential Message 005-333 to President of Chamber of Deputies, *Formula Indicación al Proyecto de Modificación del Código de Aguas*, 4 July 1996. For further explanation of the government's position, see Dirección General de Aguas 1999; Peña 1997a, 1997b.
29. Miranda 1995, D10.
30. Dirección General de Aguas 1999, 5. Alejandro Vergara, Chile's leading water law scholar, has also noted the government's change of heart about the benefits of water markets, although in my view he overstates the degree of promarket consensus and puts too little weight on the crude realities of political power, as I discuss later. See Vergara 2002, a paper presented at the fourth annual Chilean Water Law Conference, in 2001.
31. Resolution 480, *Comisión Resolutiva Anti-Monopolios*, 7 January 1997, later published in *Revista de Derecho de Aguas* 7: 285–301.
32. These issues are the subject of some of my current research, which I will present in future publications.

33. See Bauer 1998a and Bauer 1998b, 79–118, for an extended analysis of the water management issues raised by nonconsumptive water rights. A brief summary is included in Chapter 4.
34. Presidential Message 005-333 to President of Chamber of Deputies, *Formula Indicación al Proyecto de Modificación del Código de Aguas*, 4 July 1996, 1–2.
35. For arguments in favor of fees for nonuse in the previous round of debate, see ENDESA 1993; Instituto de Ingenieros 1993; Sociedad Nacional de Agricultura 1993.
36. "Las Aguas de la Discordia," *El Mercurio*, 8 September 1996, D1, D12.
37. Figueroa 1997a, A2. This lawyer had also been a vocal opponent of the first round of proposed reforms in the early 1990s; see Figueroa 1993a, 1993b.
38. Hoschild 2000, 172 (emphasis added). This paper was presented at the second annual Chilean Water Law Conference, in 1999.
39. See Domper 1996; Echeverría 1999; Figueroa 1997b; Peralta 2000; Romero 1996; "Derechos de Aguas," *El Mercurio* editorial, 6 October 1996, A3.
40. Chile has both a Constitutional Tribunal and a Supreme Court. The Supreme Court can rule on constitutional issues raised in specific cases that come before it, but only the Constitutional Tribunal has ex ante jurisdiction over constitutional issues raised by proposed legislation. See, for example, Cea 1988.
41. The deputies' challenge referred to a provision in the 1980 Constitution that guaranteed the right "*to acquire ownership of all classes of goods*, except those that Nature has made common to all men or that legislation has declared must belong to the Nation as a whole" (emphasis added). Constitution, Article 19, No. 23. This provision would apply to water rights but not to water itself, according to the legal distinction discussed in Chapter 2.
42. Constitutional Tribunal, *Requerimiento de Inconstitucionalidad del Proyecto de Ley Modificatoria del Código de Aguas*, 13 October 1997, later published in *Revista de Derecho de Aguas* 8: 299–317.
43. Jaeger 2000, 175–79. This paper was presented at the second annual Chilean Water Law Conference, in 1999. Alejandro Vergara, in contrast, strongly disagreed with the Constitutional Tribunal's decision, which he argued was an unreasonable interpretation of the meaning of Article 19, No. 23 of the Constitution. Vergara's view was that by allowing ordinary legislation to condition the granting of new water rights, the decision failed to protect the constitutional right as the Constitution's drafters had intended. See Vergara 2001, 382–87, and Note 41 above.
44. Cited in Jaeger 2000, 179–81.
45. Muñoz 1997.
46. Bertelsen 2000, 70–74. This paper was presented at the second annual Chilean Water Law Conference, in 1999.
47. The committees were first the Committee on Public Works, then the Committee on Constitution, Legislation, Justice, and Regulation, and finally the Committee on Finance. See *Informe de la Comisión de Hacienda sobre el Proyecto de Ley que Modifica el Código de Aguas*, 15 September 2000.
48. The conference was first established and has always been organized by Chile's leading water law scholar, Alejandro Vergara Blanco, who is a professor at the Catholic University Law School as well as a practicing lawyer. Conference papers and proceedings are later published in the law school's *Revista de Derecho Administrativo Económico*, of which Vergara is chief editor.
49. See, for example, Domper 1996, 2003; Figueroa 1997b; Hoschild 2000; Peralta 2000. For examples from the first round of debate about reforming the Water Code, see Confederación de Canalistas de Chile 1993a; Figueroa 1993a, 1993b; Instituto Libertad y Desar-

rollo 1993b; "Retroceso en Régimen de Aguas," editorial in *El Mercurio*, 19 February 1993, A3. For the echo in the late 1970s, see Chapter 2.

50. Figueroa 1993a, 1993b; Fromin 1998; Peralta 2000.
51. See *Informe de la Comisión de Hacienda sobre el Proyecto de Ley que modifica el Código de Aguas*, 15 September 2000; comments by Sen. Sergio Romero, Senate Session 15, *Modificación de Código de Aguas*, 5 December 2000, 49–64, and by Sen. Evelyn Matthei, ibid., 65–74; Romero 1998.
52. Gómez-Lobo and Paredes 2001, 96, 103.
53. The paper by Gómez-Lobo and Paredes makes several well-argued points that are useful contributions to the policy debate. Unfortunately, their analysis is also remarkable for being mainly an exercise in neoclassical economic theory, as the authors themselves make clear, with little basis in empirical knowledge about water rights or water markets, in Chile or elsewhere. The authors' evident lack of knowledge about the practical or institutional aspects of Chilean water markets, however, does not prevent them from weighing in with confident assertions that experience and existing empirical studies have shown to be unfounded—for example, that transaction costs are unimportant, that water rights auctions are readily feasible, and so forth.
54. Neither such a database nor all legal titles themselves currently exist, as discussed in Chapter 4.
55. See *Informe de la Comisión de Hacienda sobre el Proyecto de Ley que modifica el Código de Aguas*, 15 September 2000, 9–11; Dirección General de Aguas 1999; Jaeger 2001; Landerretche 2002.
56. Pablo Jaeger, panelist commentary, fourth annual Chilean Water Law Conference, November 2001.

Chapter 4. The Results of Chilean Water Markets

1. The relative lack of nonpartisan, empirical study is a common feature of nearly all public policy debate in Chile; it is not unique to water issues.
2. I focus on the period after 1990 because there was no research on these issues before Chile returned to democratic government, as explained at the beginning of Chapter 3.
3. World Bank 1994, ii–iii. The fact that the Chilean consultants had included these provisions suggests that they recognized some of the problems of the Chilean Water Code, although they did not say so. The Peruvian government did not approve the draft law, despite pressure from both the World Bank and the Inter-American Development Bank, because of strong opposition within Peru. See also Trawick 2003.
4. Rosegrant and Binswanger 1994, 1618–19, 1622. Their assertion that markets have reduced conflicts is unfounded, as discussed later in this chapter.
5. See Jarvis 1985, 1988.
6. See Gazmuri and Rosegrant 1994; Rosegrant and Gazmuri 1994a, 1994b. Although these papers provide some useful information, they also make confident assertions without citing any evidence, assertions that in many cases have later proved to be unfounded.
7. See Bauer 1993, 1997, 1998a, 1998b. Bauer 1998b was later published in Spanish as Bauer 2002, which was reviewed in Vergara 2003.
8. For more detailed discussion of all these factors, see Bauer 1997 and 1998b, 51–78.
9. See Chapter 2 and the references cited in the previous note.
10. See Hearne 1995; Hearne and Easter 1995; and Hearne's two chapters on Chile in Easter et al. 1998. Hearne also recognized problems caused by "the absence of institutions for

intersectoral discussion and conflict resolution" and suggested the need for a stronger government role in that area, although his own research did not address the issue further. See Hearne and Easter 1995, 40–41.

11. See Ríos and Quiroz 1995, 28–29, 15, vii. As noted in the text above, the problems they mention had been discussed in detail in Bauer 1993.
12. See Dirección General de Aguas 1999; Peña 1997a, 1997b. For a recent DGA compilation of empirical data, see Alegría et al. 2002.
13. See Solanes 1996; Solanes and Getches 1998. See Zegarra 1997 for a Peruvian economist's account of the clash between ECLAC and the Banks. Other international water experts who shared ECLAC's concerns about the exaggerated praise of the Chilean model included Luis García at the Inter-American Development Bank; Héctor Garduño at the Mexican government's National Water Commission (later a consultant for the UN Food and Agriculture Organization and the World Bank); David Getches, law professor at the University of Colorado; and the late Albert Utton, law professor at the University of New Mexico. Within the World Bank, John Briscoe also shared this concern, since in his view the exaggerated praise was counterproductive and undermined people's willingness to see that the Chilean model did indeed have great benefits (see Briscoe 1997, 17). Briscoe et al. 1998, 11, refer to the "heated debate within the World Bank" about Chilean water markets.
14. See Bauer 1997, 1998a, 1998b, and the chapters by Hearne in Easter et al. 1998.
15. See Chapter 1, Notes 42 and 46.
16. Cf. Bauer 1993, 1995. Updated and more complete versions of my research were published during this same period in the United States, but these publications were in English and were not generally available in Chile. See Bauer 1997, 1998a, 1998b.
17. The paper I summarize here is Vergara 1997b; for additional discussion along the same lines, see also Vergara 1997c. Both papers were republished in Vergara 1998, a book compiling his articles about water law written over the course of the 1990s. I also mention Vergara several times in Chapter 3.
18. Vergara 1997b, 85.
19. Vergara 1997b, 86; Peralta 1995.
20. The Water Code includes special procedures to "regularize" water rights that predated 1981, although these procedures are not mandatory. See Bauer 1998b, Chapters 3 and 4.
21. Cf. Donoso 1995 with Donoso et al. 2001.
22. Donoso 1999.
23. Donoso et al. 2001.
24. Dourojeanni and Jouravlev 1999.
25. This is the only one of the four basins studied by Hearne that had an active market, as discussed above; it is also the example that Briscoe has in mind when he says that "in well-regulated river basins in arid areas of Chile, the water markets function as one would wish," as quoted in Chapter 1. The significance of the Limarí basin was also mentioned in Bauer 1997, 645, and 1998b, 59, and it was described in various papers presented at the third and fourth National Conventions of Water Users in Chile: Confederación de Canalistas de Chile 1993b, 1997.
26. Hadjigeorgalis 1999.
27. Ibid., p.28. These institutional constraints, in turn, reflect the specific technical procedures that the local users' organizations have traditionally employed to calculate and measure annual water allotments within the different sectors of the system. If those procedures were changed, the institutional constraints would change as well.

28. Ibid., 162–66. As explained in Chapter 3, the government insists that this is an unfounded concern because the reform has been designed to exempt most farmers and to apply mainly to electricity companies and large-scale speculators.
29. Here I will summarize the 1998–1999 version of this document (Dirección General de Aguas 1999). Much of the same material can also be found in earlier papers and presentations by the head of DGA, Humberto Peña (e.g., Peña 1997a, 1997b).
30. As already quoted in Chapter 3, this statement continues, "... and as such the legal and economic system which regulates its use must encourage its efficient use by private individuals and by society as a whole. Accordingly, the principles of market economics are applicable to water resources, with the adaptations and corrections demanded by the particularities of hydrological processes." Dirección General de Aguas 1999, 5.
31. For more on the meaning of the "subsidiary state" in Chilean politics and the Chilean Constitution, see Bauer 1998b, 12–19.
32. The water problems of the urban poor in Chile—namely, access to affordable and good-quality water for drinking and domestic purposes—are not issues of water rights or water markets, and I do not discuss them here.
33. This summary is based on interviews and on the small number of available documents and publications. For further information and citations, see Bauer 1998b, 67–68. From 1992 through 2002, I interviewed staff at three Chilean nongovernmental organizations (Agraria, Grupo de Investigaciones Agrarias, and SEPADE), the Institute for Agricultural Development (part of the Ministry of Agriculture, both local offices and the central office in Santiago), and several canal users' associations. Particularly valuable contacts in Santiago have been Miguel Bahamondes, Grupo de Investigaciones Agrarias; Milka Castro, professor of anthropology, University of Chile; and Carlos Barrientos and Carmen Cancino, Institute for Agricultural Development, Ministry of Agriculture.
34. See Hadjigeorgalis 1999, 159–60, 165; Hearne and Easter 1995, 52–53.
35. See Donoso 1999.
36. Rosegrant and Binswanger 1994; World Bank 1994; and the citations in the following two footnotes.
37. See Ríos and Quiroz 1995, 27. The previous research they cite is my own publication, Bauer 1993.
38. The proponents Briscoe cites are Gazmuri and Rosegrant 1994, whose unreliability I discussed earlier in this chapter. Differences of opinion in these matters are understandable, of course, although one hopes that they have some solid basis. What is disturbing in this case is that Briscoe actually misrepresents my argument by quoting me out of context and then gives an inaccurate citation that prevents the reader from checking the source; see Briscoe 1997, 13.
39. See Note 33 above.
40. See Cancino 2001.
41. Because they take it for granted that equity has improved? Because they suspect that it has not? Somehow the latter reason seems more likely, since if these proponents could demonstrate equity improvements, they would presumably want to publicize the evidence.
42. This is one of the arguments made by the neoliberal Instituto Libertad y Desarrollo against the government's proposed Water Code reforms: María de la Luz Domper, panelist commentary, Chilean Water Law Conference, November 2001. See also Donoso 1999.
43. See, for example, Asmal 1998; Garduño 1996, 2001.
44. Dirección General de Aguas 1999.
45. See Briscoe 1996a, 22; Briscoe et al. 1998, 6–8; Hearne and Easter 1995, 40–41.

46. As I argued in Chapter 3, some of that hostile reception was due to the poor design of the proposals.
47. See Peña 2001.
48. See Bauer 1998a and 1998b, 79–118, for more detailed discussion. Bauer 1993 included an earlier analysis of these issues but was in Spanish.
49. As summarized in Chapter 2, the 1980 Constitution was intended to guarantee the permanence of the radical political, economic, and social changes imposed in Chile during the 16 years of military rule. It is the core of the military regime's institutional legacy, and the regime refused to leave office until its political opponents—today's governing coalition, the Concertación—committed themselves to that Constitution. See Chapter 2, Note 12.
50. See Bauer 1998b, 79–118, for more detailed description and complete references about these conflicts.
51. The Laja River conflict had other local ripple effects that further indicated the institutional obstacles facing river basin management. The conflict raised public awareness in the region about the need for better coordination of water uses, and during 1986–1988 the regional head of DGA and other government, business, and university leaders made attempts to create new basin organizations, loosely patterned on vigilance committees. These attempts failed for lack of legal authority and political will. See Bauer 1998b, 90–91.
52. Water Code, Article 14. Articles 15 and 97 also place restrictions on nonconsumptive rights in particular circumstances, but Article 14 is the most important; see Bauer 1998b, 84.
53. Before 1981 there was no category of "consumptive" water rights because it was taken for granted that *all* water rights were consumptive, reflecting the historical predominance of irrigation in Chile. Under previous laws there were other administrative means of recognizing the distinctive water rights needs of power generators. This raises another problem of interpretation for pre-1981 hydroelectric water rights: should they be considered consumptive or nonconsumptive?
54. This bias in voting rights appears to have been an inadvertent oversight on the part of the code's drafters rather than a deliberate decision, although it is hard to know for certain.
55. See also Note 53 above.
56. Colbún is in fact two adjoining dams and reservoirs that are managed jointly; the second reservoir, Machicura, is smaller and serves to regulate the flows being discharged back to the river channel. I will refer to both dams and reservoirs jointly as Colbún for the sake of simplicity.
57. For a blow-by-blow description, see Bauer 1998b, 94–110.
58. The more important structure for regulating the Bío Bío River's flow is ENDESA's Ralco Dam, to be built later and farther upstream, with much greater storage capacity. Ralco has been as controversial as Pangue and was still under construction at the time of this book's publication.
59. Barzel 2001, personal communication. See Barzel 1989 for a general analysis of the economics of property rights, particularly 89–91 on water rights.
60. For an analysis of all published court decisions from 1981 through 1993, apart from the cases involving the Maule and Bío Bío River basins, see Bauer 1998b, 80–84. My subsequent research on all published court decisions through 2001 confirms the general tendencies discussed here but has not yet been published.
61. DGA's tendency toward strict legalism has been reinforced by the close scrutiny by the Office of the Controller General, an autonomous government agency responsible for auditing the legal and administrative actions of other government agencies and enterprises. Under Director Peña, DGA has resisted strong political pressures to grant new water rights in certain situations, namely, ENDESA's applications for additional nonconsumptive rights (see Chapter 3, Note 31, and accompanying text) and the exploding

demand for new groundwater rights in northern Chile (see the section, "Emerging Issues," at the end of this chapter). The agency's behavior, however, has been a defense of one of its core areas of authority—the constitution of new property rights—rather than an aggressive move toward increased regulation.

62. Bauer 1993, 1998a, 1998b, 1998c.
63. Donoso 1999, 307.
64. The notable exception is Alejandro Vergara, who despite being an orthodox Chilean lawyer and legal scholar in most respects, has exerted himself to understand the language of other disciplines and addressed the practical issues of water rights. See, for example, Vergara 1997b, 1998.
65. This list reflects my own observations based on three sources of information. First, I have made research trips to Chile every year from 1995 to 2003, for periods from several weeks to several months, during which I have done interviews with both established and new professional contacts. Second, I have been a regular reader of Chilean newspapers and weekly news magazines (in print and via the Internet) during that same period. Third, I have participated in all the annual Chilean Water Law Conferences held to date, beginning in 1998. The proceedings of these conferences are an excellent source of information and have been published in the Chilean *Journal of Administrative and Economic Law* (*Revista de Derecho Administrativo Económico*).

Chapter 5. Conclusions and Lessons about the Chilean Experience

1. See Chapter 4 and citations therein for more detail.
2. See Aguilera and Sánchez 2002 for a recent application of this argument to water markets in the Canary Islands of Spain.
3. See Chapter 4. Humberto Peña, who has been the head of DGA since 1994, made a similar assessment about the Water Code's strengths and weaknesses in his own review of the code's first 20 years. He said that the code was strongest in its treatment of the economic aspects of water and weakest in environmental and social aspects, and he emphasized that IWRM was simply not considered in 1981. See Peña 2001.
4. See Chapter 3 and citations therein for more detail.
5. Some officials in DGA insist that progress has indeed been made despite the lack of legislative change. They argue that the years of debate have convinced most people that the government's diagnosis of the Water Code's flaws is correct, even though people may disagree about the government's proposed remedy. They also point to the favorable decisions by the Constitutional Tribunal and the Anti-Monopoly Commission in 1997, cases that many observers expected the government to lose, as significant victories for the government's position (see Chapter 3). Interview with Humberto Peña, director, and Pablo Jaeger, chief lawyer, Dirección General de Aguas, 29 May 2002.
6. For an example of this reasonable tendency in the opposition, see Sen. Sergio Romero's presentation in a seminar held in the Senate to discuss the proposed Water Code reforms (Romero 1998).
7. In his review of the first 20 years' experience of the Water Code, the head of DGA underlines the need to begin a process of "deep reflection about the forms of conflict resolution considered in the Code" (Peña 2001, 12).
8. See Chapter 1, Notes 46–50, and accompanying text. One example is a World Bank paper arguing that, despite the problems mentioned, the Chilean "system of tradable water

rights and associated water markets is a great achievement and is universally agreed to be the bedrock on which to refine Chilean water management practices" (Briscoe et al. 1998, 9). Another example is a paper for the Global Water Partnership describing Chile as "a world leader in water governance." The authors argue that although "many mistakes with openness, transparency, participation, and ecosystem concerns were made in the hurry to get effective water markets established.... The system is adaptive and now these concerns are being addressed 20 years after the initial laws were passed" (Rogers and Hall 2002, 25–26).

9. According to John Briscoe at the World Bank, lead author of one of the papers cited in the previous note, Chile is a model of international "good practice" in managing water scarcity but not in managing water quality (see Briscoe 1996a, 20–22). That description is revealing about the limitations of the Chilean model for integrated water resources management, since a basic principle of IWRM is that water quantity and water quality are intimately related and should not be managed by different institutional frameworks. Briscoe does not point this out, however.

10. See Chapter 1, Notes 34–39, and accompanying text.

11. See Introduction, Note 1, and accompanying text.

REFERENCES

Aguilera, Federico. 1998. Hacia Una Nueva Economía del Agua: Cuestiones Fundamentales. In *El Agua a Debate Desde la Universidad: Hacia una Nueva Cultura del Agua* (Congreso Ibérico sobre Gestión y Planificación de Aguas, September, Zaragoza, Spain), edited by Pedro Arrojo and Javier Martínez. First ed. Zaragoza, Spain: Navarro & Navarro, 15–31.

Aguilera, Federico, Eduardo Pérez, and Juan Sánchez. 2000. The Social Construction of Scarcity: The Case of Water in Tenerife (Canary Islands). *Ecological Economics* 34: 233–45.

Aguilera, Federico, and Miguel Sánchez. 2002. *Los Mercados de Agua en Tenerife*. Bilbao, Spain: Bakeaz.

Alegría, María Angélica, Fernando Valdés, and Adrián Lillo. 2002. El Mercado de Aguas: Análisis Teórico y Empírico. *Revista de Derecho Administrativo Económico* IV(1): 169–85.

Angell, Alan. 1993. *Chile de Alessandri a Pinochet: En Busca de la Utopía*. Santiago: Editorial Andrés Bello.

Anguita, Pablo. 1995. *Gestión de Recursos Hídricos: Propuesta*, Irrigation Directorate, Ministry of Public Works. Unpublished document, June.

Arrojo, Pedro. 1995. Del Estructuralismo Hidráulico a la Economía Ecológica del Agua. *Mientras Tanto* 62: 77–105.

Arrojo, Pedro, and José Manuel Naredo. 1997. *La Gestión del Agua en España y California.* Bilbao, Spain: Bakeaz.

Asenjo, Rafael. 1990. Políticas Gubernamentales Sobre Protección del Medio Ambiente. In *Protección del Medio Ambiente,* edited by Herman Schwember. Proceedings of Seminar, Asociación de Ingenieros Consultores de Chile/Tecniberia. Santiago, 22–26.

Asmal, Kader. 1998. Water, Life and Justice: A Late 20th Century Reflection from the South. *The 1998 Abel Wolman Distinguished Lecture,* 21 May 1998. Water Science and Technology Board of the National Research Council (ed.), Washington, DC: National Academy of Sciences.

Bardhan, Pranab. 1989. Alternative Approaches to the Theory of Institutions in Economic Development. In *The Economic Theory of Agrarian Institutions,* edited by Pranab Bardhan. Oxford: Clarendon Press, 3–17.

Barlow, Maude. 2000. Commodification of Water—Wrong Prescription. Presentation to 10th Stockholm Water Symposium, Stockholm, 17 August 2000. http://www.canadians.org/blueplanet/pubs-barlow-speech.html (accessed 18 January 2001).

Barros, Robert. 2002. *Constitutionalism and Dictatorship: Pinochet, the Junta, and the 1980 Constitution.* New York: Cambridge University Press.

Barzel, Yoram. 1989. *Economic Analysis of Property Rights.* Cambridge: Cambridge University Press.

Bates, Sarah, David Getches, Lawrence MacDonnell, and Charles Wilkinson. 1993. *Searching Out the Headwaters: Change and Rediscovery in Western Water Policy.* Washington, DC: Island Press.

Bauer, Carl. 1993. Los Derechos de Agua y el Mercado: Efectos e Implicancias del Código de Aguas Chileno de 1981. *Revista de Derecho de Aguas* 4: 17–63.

———. 1995. Against the Current? Privatization, Market and the State in Water Rights: Chile 1979–1993. Ph.D. dissertation, Jurisprudence and Social Policy Program, University of California–Berkeley.

———. 1996. El Mercado de Aguas en California. In *Precios y Mercados del Agua,* edited by Antonio Embid. Madrid: Editorial Civitas, 179–205.

———. 1997. Bringing Water Markets Down to Earth: The Political Economy of Water Rights in Chile, 1976–1995. *World Development* 25(5): 639–56.

———. 1998a. Slippery Property Rights: Multiple Water Uses and the Neoliberal Model in Chile, 1981–1995. *Natural Resources Journal* 38(1): 109–155.

———. 1998b. *Against the Current: Privatization, Water Markets, and the State in Chile.* Boston: Kluwer Academic Publishers.

———. 1998c. Derecho y Economía en la Constitución de 1980. *Perspectivas en Política, Economía y Gestión* 2(1): 23–47.

———. 2002. *Contra la Corriente: Privatización, Mercados de Agua y el Estado en Chile.* Santiago: LOM Ediciones.

Benda-Beckmann, F. von, K. von Benda-Beckmann, and H.L. Joep Spiertz. 1997. Local Law and Customary Practices in the Study of Water Rights. In *Water Rights, Conflict and Policy,* edited by Rajendra Pradhan et al. Proceedings of workshop held in Kathmandu, Nepal, January 1996. Colombo, Sri Lanka: International Irrigation Management Institute, 221–42.

Bertelsen, Raúl. 2000. Análisis Constitucional de la Reforma del Código de Aguas. *Revista de Derecho Administrativo Económico* II(1): 63–74.

Birdsall, Nancy, Carol Graham, and Richard Sabot. 1998. Virtuous Circles in Latin America's Second Stage of Reforms. In *Beyond Tradeoffs: Market Reforms and Equitable Growth in Latin America,* edited by Nancy Birdsall, Carol Graham, and Richard Sabot. Washington, DC: Inter-American Development Bank and Brookings Institution Press, 1–27.

Briscoe, John. 1996a. Water as an Economic Good: The Idea and What It Means in Practice. Paper presented at World Congress of the International Commission on Irrigation and Drainage, September, Cairo.

———. 1996b. *Water Resources Management in Chile: Lessons from a World Bank Study Tour.* Washington, DC: World Bank.

———. 1997. Managing Water as an Economic Good: Rules for Reformers. *Water Supply* 15(4): 153–72.

Briscoe, John, Pablo Anguita, and Humberto Peña. 1998. *Managing Water as an Economic Resource: Reflections on the Chilean Experience.* Environment Department Paper No. 62. Washington, DC: World Bank.

Bromley, Daniel. 1982. Land and Water Problems: An Institutional Perspective. *American Journal Agricultural Economics* December: 834–44.

———. 1989. *Economic Interests and Institutions: The Conceptual Foundations of Public Policy.* New York: Basil Blackwell.

———. 1991. *Environment and Economy: Property Rights and Public Policy.* Cambridge, MA: Basil Blackwell.

Brown, F. Lee. 1997. Water Markets and Traditional Water Values: Merging Commodity and Community Perspectives. *Water International* 22(1): 2–5.

Brown, F. Lee, and Helen Ingram. 1987. *Water and Poverty in the Southwest.* Tucson, AZ: University of Arizona Press.

Brown, F. Lee, et al. 1982. Water Reallocation, Market Proficiency, and Conflicting Social Values. In *Water and Agriculture in the Western U.S.: Conservation, Reallocation, and Markets,* edited by Gary Weatherford. Boulder, CO: Westview Press, 191–256.

Bruns, Bryan, and Ruth Meinzen-Dick (eds.). 2000. *Negotiating Water Rights.* New Delhi: Vistaar Publications/International Food Policy Research Institute.

Cancino, Carmen. 2001. Proyecto Piloto: Conflicto de Aguas de Regadío en el Sector Las Pataguas–Valdivia de Paine. Paper presented at third Encuentro de las

Aguas, Instituto Interamericano de Cooperación para la Agricultura, October, Santiago.

Cavallo, Ascanio, Manuel Salazár, and Oscar Sepúlveda. 1989. *La Historia Oculta del Régimen Militar*. Santiago: Editorial Antártica.

Cea, José Luis. 1988. *Tratado de la Constitución de 1980: Características Generales, Garantías Constitucionales*. Santiago: Editorial Jurídica de Chile.

Challen, Roy. 2001. Economic Analysis of Alternative Institutional Structures for Governance of Water Use. Paper presented at 45th Annual Conference of Australian Agricultural and Resource Economics Society, 23–25 January, Adelaide, Australia.

Ciriacy-Wantrup, S.V. 1967. Water Economics: Relations to Law and Policy. In *Waters and Water Rights*, edited by Robert E. Clark. Indianapolis: Allen Smith, 397–430.

Clark, Sandford. 1999. Reforming South African Water Legislation: Tradable Water Entitlements in Australia. In *Issues in Water Law Reform*, edited by Stefano Burchi. Rome: UN Food and Agriculture Organization, 23–51.

Coase, Ronald. 1988. *The Firm, the Market, and the Law*. Chicago: University of Chicago Press.

Cole, Daniel, and Peter Grossman. 2002. The Meaning of Property Rights: Law versus Economics? *Land Economics* 78(3): 317–30.

Commons, John. 1924. *The Legal Foundations of Capitalism*. New York: Macmillan.

———. 1934. *Institutional Economics*. New York: Macmillan.

Confederación de Canalistas de Chile. 1993a. *Comentario a las Modificaciones del Código de Aguas*. Unpublished document.

———. 1993b. Tercera Convención Nacional de Regantes de Chile. Los Angeles, Chile, November.

———. 1997. Cuarta Convención Nacional de Usuarios del Agua. Arica, Chile, October.

Constable, Pamela, and Arturo Valenzuela. 1991. *A Nation of Enemies: Chile under Pinochet*. New York: W.W. Norton.

Cosgrove, William, and Frank Rijsberman. 2000. *World Water Vision: Making Water Everybody's Business*. London: World Water Council.

Costanza, Robert, Olman Segura, and Juan Martínez-Alier (eds.). 1996. *Getting Down to Earth: Practical Applications of Ecological Economics*. Washington, DC: Island Press.

Dakolias, Maria. 1996. The Judicial Sector in Latin America and the Caribbean: Elements of Reform. Technical Paper No. 319. Washington, DC: World Bank.

Daly, Herman, and Kenneth Townsend. 1993. *Valuing the Earth: Economics, Ecology, Ethics*. Cambridge, MA: MIT Press.

Del Moral, Leandro, and David Saurí. 1999. Changing Course: Water Policy in Spain. *Environment* (July-August): 12–36.

Dirección General de Aguas. 1991a. Bases para la Formulación de la Política Nacional de Aguas. Paper presented at Seminario sobre Política Nacional de Aguas, Comisión Económica para América Latina y el Caribe (CEPAL). August, Santiago. Published in *Revista de Derecho de Minas y Aguas* 2: 259–65.

———. 1991b. Minuta de Modificaciones al Código de Aguas: Conceptos Básicos a Desarrollar. *Revista de Derecho de Minas y Aguas* 2: 267–69.

———. 1999. *Política Nacional de Recursos Hídricos*. Santiago: Ministerio de Obras Públicas.

Domper, María de la Luz. 1996. Propiedad sobre el Agua. *El Mercurio*, 1 December, A2.

———. 2003. La Eficiencia en el Mercado de Derechos de Agua: ¿Patente por No Uso o por Tenencia? *Serie Informe Económico* No. 141, Instituto Libertad y Desarrollo. Santiago.

Donoso, Guillermo. 1994. Proyecto de Reforma al Código de Aguas: ¿Mejora la Asignación del Recurso? *Panorama Económico de la Agricultura* 16(92): 4–11.

———. 1995. El Mercado de Derechos de Aprovechamiento como Mecanismo Asignador del Recurso Hídrico. *Revista de Derecho de Aguas* 6: 9–18.

———. 1999. Análisis del Funcionamiento del Mercado de los Derechos de Aprovechamiento de Agua e Identificación de sus Problemas. *Revista de Derecho Administrativo Económico* I(2): 295–314.

———. 2000. Tarificación: ¿Es una Reforma Aplicable para Mejorar la Eficiencia de la Asignación? *Revista de Derecho Administrativo Económico* II(1): 113–20.

Donoso, Guillermo, Juan Pablo Montero, and Sebastián Vicuña. 2001. Análisis de los Mercados de Derechos de Aprovechamiento de Agua en las Cuencas del Maipo y el Sistema Paloma en Chile: Efectos de la Variabilidad de la Oferta Hídrica y de los Costos de Transacción. *Revista de Derecho Administrativo Económico de Recursos Naturales* III(2): 367–87.

Dourojeanni, Axel. 1994. Water Management and River Basins in Latin America. *CEPAL Review* 53: 111–28.

Dourojeanni, Axel, and Andrei Jouravlev. 1999. *El Código de Aguas de Chile: Entre la Ideología y la Realidad*. Serie Recursos Naturales e Infraestructura No.3, División de Recursos Naturales e Infraestructura, Comisión Económica Para América Latina. Santiago: Naciones Unidas.

Drake, Paul, and Ivan Jaksic (eds.). 1991. *The Struggle for Democracy in Chile, 1982–90*. Lincoln, NE: University of Nebraska Press.

Dubash, Navroz. 2002. *Tubewell Capitalism: Groundwater Development and Agrarian Change in Gujarat*. Oxford: Oxford University Press.

Easter, K. William, Mark Rosegrant, and Ariel Dinar (eds.). 1998. *Markets for Water: Potential and Performance*. Boston: Kluwer Academic Publishers.

Echeverría, Germán. 1999. La Pelea que Viene por el Agua. *El Mercurio*, 16 December, A1, A11.

Ellenberg, Jorge. 1980. *Antecedentes Respecto del Nuevo Régimen Legal de Aguas*. Tesis, Facultad de Derecho, Universidad de Chile.

Embid, Antonio (ed.). 1996. *Precios y Mercados del Agua*. Madrid: Editorial Civitas.

ENDESA. 1993. *El Derecho de Aprovechamiento de Agua en Chile: Visión de ENDESA*. Unpublished document.

Field, Alexander. 1981. The Problem with Neoclassical Institutional Economics: A Critique with Special Reference to the North/Thomas Model of pre-1500 Europe. *Explorations in Economic History* 18: 174–98.

Figueroa, Luis Simón. 1989. La Asignación de los Derechos de Aprovechamiento de Aguas, un Debate Pendiente. Paper presented at second Convención Nacional de Regantes de Chile, September, La Serena, Chile.

———. 1993a. Estatuto Jurídico de las Aguas: Evolución Histórica y Cultural. *Derecho en la Región* 1: 25–36.

———. 1993b. Cambios a la Legislación de Aguas. *El Mercurio*, 27 January, A2.

———. 1997a. La Planificación al Ataque. *El Mercurio*, 5 June, A2.

———. 1997b. Consecuencias Prácticas de las Modificaciones al Código de Aguas. Paper presented at fourth Convención Nacional de Usuarios del Agua. October, Arica, Chile. Santiago: Confederación de Canalistas de Chile, 101–17.

Fontaine, Arturo. 1988. *Los Economistas y el Presidente Pinochet*. Second ed. Santiago: Zig Zag.

Food and Agriculture Organization. 1999. *Issues in Water Law Reform*. FAO Legislative Study 67. Rome.

Food and Agriculture Organization, UN Development Programme, and World Bank. 1995. *Water Sector Policy Review and Strategy Formulation: A General Framework*. FAO Land and Water Bulletin 3. Rome.

Fort, Denise. 1999. The Western Water Commission: Watershed Management Receives the Attention of a New Generation. *Journal of the American Water Resources Association* 35(2): 223–32.

Frederick, Kenneth (ed.). 1986. *Scarce Water and Institutional Change*. Washington, DC: Resources for the Future.

Fromin, Luis. 1998. Se Agita la 'Liquidez'. *El Mercurio*, 13 September, B6–7.

García, Luis. 1998. *Integrated Water Resources Management in Latin America and the Caribbean*. Washington, DC: Inter-American Development Bank.

García, Sergio. 1999. Un Análisis Crítico de la Política Ambiental de la Concertación. *Perspectivas en Política, Economía y Gestión* 3(1): 163–89.

Garduño, Héctor. 1996. *Water Use Management in Mexico: Strategy for the 1995–2000 Period*. Texas Natural Resource Conservation Commission and Mexican National Water Commission.

———. 2001. *Water Rights Administration: Experience, Issues and Guidelines*. FAO Legislative Study 70. Rome.

Garrido, José, Cristián Guerrero, and María Soledad Valdés. 1990. *Historia de la Reforma Agraria en Chile*. Second ed. Santiago: Editorial Universitaria.

Gazmuri, Renato, and Mark Rosegrant. 1994. Chilean Water Policy: The Role of Water Rights, Institutions, and Markets. In *Tradable Water Rights: Experiences in Reforming Water Allocation Policy*, by Mark Rosegrant and Renato Gazmuri, document prepared for U.S. Agency for International Development, Irrigation Support Project for Asia and the Near East.

Getches, David. 1996. Changing the River's Course: Western Water Policy Reform. *Environmental Law* 26: 157.

Gleick, Peter (ed.). 1997. *Water in Crisis: A Guide to the World's Freshwater Resources*. Oxford: Oxford University Press.

———. 1998. *The World's Water 1998–1999: The Biennial Report on Freshwater Resources*. Washington, DC: Island Press.

Gleick, Peter, et al. 2002. *The New Economy of Water: The Risks and Benefits of Globalization and Privatization of Fresh Water*. Oakland, CA: Pacific Institute.

Global Water Partnership. 1998. *A Strategic Plan: Global Water Partnership 1999*. Stockholm: Global Water Partnership.

———. 2000a. *Towards Water Security: A Framework for Action*. Stockholm: Global Water Partnership.

———. 2000b. *Integrated Water Resources Management*. Technical Advisory Committee Background Paper No. 4. Stockholm: Global Water Partnership.

Gómez-Lobo, Andrés, and Ricardo Paredes. 2001. Mercado de Derechos de Aguas: Reflexiones sobre el Proyecto de Modificación del Código de Aguas. *Revista de Estudios Públicos* 82 (Autumn): 83–104.

Graham, Carol, and Moisés Naím. 1998. The Political Economy of Institutional Reform in Latin America. In *Beyond Tradeoffs: Market Reforms and Equitable Growth in Latin America*, edited by Nancy Birdsall, Carol Graham, and Richard Sabot. Washington, DC: Inter-American Development Bank and Brookings Institution Press, 321–61.

Hadjigeorgalis, Ereney. 1999. *Private Water Markets in Agriculture and the Effects of Risk, Uncertainty and Institutional Constraints*. Ph.D. dissertation, Department of Agricultural Economics, University of California–Davis.

Hearne, Robert. 1995. *The Market Allocation of Natural Resources: Transactions of Water Use Rights in Chile*. Ph.D. dissertation, Department of Agricultural Economics, University of Minnesota.

Hearne, Robert, and K. William Easter. 1995. *Water Allocation and Water Markets: An Analysis of Gains-from-Trade in Chile*. Technical Paper No. 315. Washington, DC: World Bank.

Hildreth, Richard G. 1999. Water Law at the Crossroads. *Journal of Environmental Law and Litigation* 14(1): 1.

Hodgson, Geoffrey. 1988. *Economics and Institutions: A Manifesto for a Modern Institutional Economics*. Philadelphia: University of Pennsylvania Press.

Hoschild, Hernan. 2000. Posición del Empresario Minero Frente a la Reforma del Código de Aguas. *Revista de Derecho Administrativo Económico* II(1): 169–74.

Howe, Charles. 2000. Protecting Public Values in a Water Market Setting: Improving Water Markets to Increase Economic Efficiency and Equity. *University of Denver Water Law Review* 3(2).

Instituto de Ingenieros. 1993. *Proyecto de Ley que Modifica el Código de Aguas.* Unpublished document.

Instituto Libertad y Desarrollo. 1993a. *Análisis y Comentario de la Propuesta Reforma del Código de Aguas.* Boletín 876-09. Unpublished document.

———. 1993b. Peligrosa Vuelta Atrás en la Legislación de Aguas. *El País*, 28 October, 7.

———. 2003. Reforma al Código de Aguas: Otro paso Atrás. Temas Públicos No. 624. Santiago.

Inter-American Development Bank. 1999. Second Generation Issues in the Reform of Public Services. Conference, 4–5 October 1999, Washington, DC.

International Conference on Water and the Environment. 1992. The Dublin Statement and Report of the Conference. *International Conference on Water and the Environment: Development Issues for the 21st Century.* 26–31 January 1992, Dublin.

Jaeger, Pablo. 2000. Aspectos Relevantes de la Tramitación Parlamentaria de la Modificación al Código de Aguas. *Revista de Derecho Administrativo Económico* II(1): 175–88.

———. 2001. El Proyecto de Modificación del Código de Aguas: Avances Recientes y Perspectivas. Paper presented at fourth Jornadas de Derecho de Aguas, Facultad de Derecho, Pontificia Universidad Católica de Chile. 19–20 November, Santiago.

Jarvis, Lovell. 1985. *Chilean Agriculture under Military Rule: From Reform to Reaction, 1973–80.* Berkeley, CA: University of California–Berkeley Institute for International Studies.

———. 1988. The Unraveling of Chile's Agrarian Reform, 1973–86. In *Searching for Agrarian Reform in Latin America*, edited by W. Thiesenhausen. Winchester, MA: Unwin Hyman, 240–75.

Kapur, Devesh, John Lewis, and Richard Webb (eds.). 1997. *The World Bank: Its First Half Century.* Washington, DC: Brookings Institution Press.

Lagos, Ricardo. 1994. Discurso en el Aniversario de la Dirección General de Aguas. *Revista de Derecho de Aguas* 5: 123–28.

Landerretche, Oscar. 2002. Consideraciones Económicas sobre la Reforma del Código de Aguas. *Revista de Derecho Administrativo Económico* IV(1): 303–15.

Libecap, Gary. 1989. *Contracting for Property Rights.* Cambridge, UK: Cambridge University Press.

Livingston, Marie. 1993a. Normative and Positive Aspects of Institutional Economics: The Implications for Water Policy. *Water Resources Research* 29(4): 815–21.

———. 1993b. *Designing Water Institutions: Market Failures and Institutional Response.* Washington, DC: World Bank.

Lord, William, and Morris Israel. 1996. *A Proposed Strategy to Encourage and Facilitate Improved Water Resources Management in Latin America and the Caribbean*. March. Washington, DC: Environment Division, Social Programs and Sustainable Development Department, Inter-American Development Bank.

Loveman, Brian. 1988. *Chile: The Legacy of Hispanic Capitalism*. New York: Oxford University Press.

Manríquez, Gustavo. 1993. Política Nacional de Aguas: Formulación, Objetivos, Instrumentos, Opciones, Alternativas, y Proposiciones. *Derecho en la Región* 1: 65–80.

McNeill, Desmond. 1998. Water as an Economic Good. *Natural Resources Forum* 22(4): 253–61.

Meinzen-Dick, Ruth. 1996. Groundwater Markets in Pakistan: Participation and Productivity. International Food Policy Research Institute Research Report No. 105. Washington, DC: IFPRI.

Mercuro, Nicholas, and Steven Medema. 1997. *Economics and the Law: From Posner to Post-Modernism*. Princeton, NJ: Princeton University Press.

Miller, Kathleen, Steven Rhodes, and Lawrence MacDonnell. 1996. Global Change in Microcosm: The Case of U.S. Water Institutions. *Policy Sciences* 29: 271–90.

———. 1997. Water Allocation in a Changing Climate: Institutions and Adaptation. *Climatic Change* 35: 157–77.

Miranda, Soledad. 1995. ¿Quién Es el Dueño de las Aguas? *El Mercurio*, 23 April, D9-D10.

Muñoz, Gonzalo. 1997. Informe del Proyecto de Ley de Modificación del Código de Aguas, sobre Patentes a los Derechos de Aprovechamiento de Aguas y Otras Materias. *Revista de Derecho de Aguas* 8: 93–107.

Muñoz, Jaime. 1991. Política Nacional de Aguas. Paper presented at Jornadas sobre Uso y Conservación de Recursos Hídricos, Comité Chileno para el Programa Hidrológico Internacional. 21–24 August, La Serena, Chile.

North, Douglass. 1981. *Structure and Change in Economic History*. New York: Norton.

Peña, Humberto. 1997a. *Modificaciones al Código de Aguas y su Aporte a la Gestión del Agua*. Santiago de Chile: Dirección General de Aguas, Ministerio de Obras Públicas.

———. 1997b. Exposición del Director General de Aguas. Paper presented at fourth Convención Nacional de Usuarios del Agua. 17–18 October, Arica, Chile. Santiago: Confederación de Canalistas de Chile, 119–38.

———. 2001. 20 Años del Código de Aguas: Visión Desde la Administración. Paper presented at fourth Jornadas de Derecho de Aguas, Facultad de Derecho, Pontificia Universidad Católica de Chile. 19–20 November, Santiago.

Peralta, Fernando. 1995. El Mercado de Aguas en Chile. Paper presented at Workshop on Issues in Privatization of Water Utilities in the Americas, American Society of Civil Engineers and Economic Commission for Latin American and the Caribbean, 4–6 October, Santiago.

———. 2000. Hacia una Política de Recursos Hídricos en Chile. *Revista de Derecho Administrativo Económico* II(1): 253–59.

Perry, C.J., David Seckler, and Michael Rock. 1997. *"Water as an Economic Good": A Solution, or a Problem?* Research Report No. 14. Colombo, Sri Lanka: International Irrigation Management Institute, Winrock International Institute for Agricultural Development.

Prugh, Thomas, et al. 1999. *Natural Capital and Human Economic Survival*. Lewis Publishers, Inc.

Public Services International. 2000. Controlling the Vision and Fixing the Forum: The Politburo of Privatization. Briefing paper. Public Services International Research Unit, University of Greenwich, London.

Ríos, Mónica, and Jorge Quiroz. 1995. *The Market of Water Rights in Chile: Major Issues*. Technical Paper No. 285. Washington, DC: World Bank.

Rogers, Peter. 2002. Water Governance. Draft paper presented at annual meeting, Inter-American Development Bank, March, Fortaleza, Brazil.

Rogers, Peter, Ramesh Bhatia, and Annette Huber. 1998. *Water as Social and Economic Good: How to Put the Principle into Practice*. Stockholm: Global Water Partnership Technical Advisory Committee.

Rogers, Peter, and Alan Hall. 2002. *Effective Water Governance*. Stockholm: Global Water Partnership.

Romero, Sergio. 1996. Dejar Abierta la Llave del Agua? *El Mercurio*, 22 October, A2.

———. 1998. Panorama General del Mercado de Aguas y Equidad y Eficiencia en la Asignación Originaria de Derechos de Aguas. *Revista Derecho de Aguas* IX: 321–28.

Rosegrant, Mark, and Hans Binswanger. 1994. Markets in Tradable Water Rights: Potential for Efficiency Gains in Developing Country Water Resource Allocation. *World Development* 22: 1613–25.

Rosegrant, Mark, and Renato Gazmuri. 1994a. Tradable Water Rights: Experiences in Reforming Water Allocation Policy. U.S. Agency for International Development. Irrigation Support Project for Asia and the Near East.

———. 1994b. Reforming Water Allocation Policy through Markets in Tradable Water Rights: Lessons from Chile, Mexico, and California. Environment and Production Technology Division Discussion Paper No. 6. Washington, DC: International Food Policy Research Institute.

Rowat, Malcolm, et al. 1995. Judicial Reform in Latin America and the Caribbean: Proceedings of a World Bank Conference. Technical Paper No. 280. Washington, DC: World Bank.

Ruthenberg, Ina-Marlene, et al. 2001. A Decade of Environmental Management in Chile. Environment Department Paper No. 82. Washington, DC: World Bank.

Rutherford, Malcolm. 2001. Institutional Economics: Then and Now. *Journal of Economic Perspectives* 15(3): 173–94.

Saliba, Bonnie Colby, and David Bush. 1987. *Water Markets in Theory and Practice: Market Transfers, Water Values, and Public Policy*. Studies in Water Policy and Management No. 12. Boulder, CO: Westview Press.

Sax, Joseph, Robert Abrams, and Barton Thompson, Jr. 1991. *Legal Control of Water Resources: Cases and Materials*. Second edition. St. Paul: West Publishing Co.

Shah, Tushar. 1993. *Groundwater Markets and Irrigation Development: Political Economy and Practical Policy*. Bombay: Oxford University Press.

Shupe, Steven, Gary Weatherford, and Elizabeth Checchio. 1989. Western Water Rights: The Era of Reallocation. *Natural Resources Journal* 29: 413–34.

Silva, Eduardo. 1994. Contemporary Environmental Politics in Chile: The Struggle over the Comprehensive Law. *Industrial and Environmental Crisis Quarterly* 8(4): 323–43.

———. 1997. Chile. In *National Environmental Policies: A Comparative Study of Capacity-Building*, edited by Martin Janicke and Helmut Weidner. Berlin: Springer, 213–33.

Simpson, Larry, and Klas Ringskog. 1997. *Water Markets in the Americas*. Washington, DC: World Bank.

Sociedad Nacional de Agricultura. 1993. Código de Aguas: Observaciones de la SNA al Proyecto que Modifica la Ley. *El Campesino* (Junio): 8–14.

Solanes, Miguel. 1996. Mercados de Derechos de Agua: Componentes Institucionales. *Revista de la CEPAL* 59: 83–96.

———. 1998. Manejo Integrado del Recurso del Agua, con la Perspectiva de los Principios de Dublin. *Revista de la CEPAL* 64: 165–85.

Solanes, Miguel, and David Getches. 1998. Prácticas Recomendables para la Elaboración de Leyes y Regulaciones Relacionadas con el Recurso Hídrico. *Informe de Buenas Prácticas* No. ENV-127. Washington, DC: Banco Interamericano de Desarrollo.

Solanes, Miguel, and Fernando González. 1999. *The Dublin Principles for Water as Reflected in a Comparative Assessment of Institutional and Legal Arrangements for Integrated Water Resources Management*. Stockholm: Global Water Partnership Technical Advisory Committee.

Tarlock, A. Dan. 1991. New Water Transfer Restrictions: The West Returns to Riparianism. *Water Resources Research* 27(6): 987–94.

Trawick, Paul. 2003. Against the Privatization of Water: An Indigenous Model for Improving Existing Laws and Successfully Governing the Commons. *World Development* 31(6): 977–96.

Urquidi, Juan Carlos. 1994. Análisis Crítico de la Institucionalidad y del Marco Regulatorio del Recurso Hídrico Continental. *Revista de Derecho de Aguas* 5: 61–79.

Valenzuela, Arturo. 1989. Chile: Origins, Consolidation, and Breakdown of a Democratic Regime. In *Democracy in Developing Countries, Vol. 4: Latin America*, edited by L. Diamond et al. Boulder, CO: Lynne Rienner Publishers, 159–206.

———. 1991. The Military in Power: The Consolidation of One-Man Rule. In *The Struggle for Democracy in Chile, 1982–90*, edited by Paul Drake and Ivan Jaksic. Lincoln, NE: University of Nebraska Press, 21–72.

Vergara, Alejandro. 1992. *Principios y Sistema del Derecho Minero: Estudio Histórico-Dogmático*. Santiago: Editorial Jurídica de Chile.

———. 1997a. El Catastro Público de Aguas. Consagración Legal, Contenido y Posibilidades de Regulación Reglamentaria. *Revista de Derecho de Aguas* 8: 71–91.

———. 1997b. Perfeccionamiento Legal del Mercado de Derecho de Aprovechamiento de Aguas. Paper presented at fourth Convención Nacional de Usuarios del Agua. 17–18 October, Arica, Chile. Santiago: Confederación de Canalistas de Chile, 83–96.

———. 1997c. La Libre Transferibilidad de los Derechos de Agua: El Caso Chileno. *Revista Chilena de Derecho* 24(2): 369–95.

———. 1998. *Derecho de Aguas*. Santiago: Editorial Jurídica de Chile.

———. 2001. La *Summa Divisio* de Bienes y Recursos Naturales en la Constitución de 1980. In *Veinte Años de la Constitución Chilena: 1981–2001*. Santiago: Ediar Conosur, 369–89.

———. 2002. Las Aguas como Bien Público (No Estatal) y lo Privado en el Derecho Chileno: Evolución Legislativa y su Proyecto de Reforma. *Revista de Derecho Administrativo Económico* IV(1): 63–79.

———. 2003. Review of *Contra la Corriente: Privatización, Mercados de Agua y el Estado en Chile*, by Carl Bauer. *Revista Chilena de Derecho* 30(1): 409–19.

Wandschneider, Philip. 1986. Neoclassical and Institutionalist Explanations of Changes in Northwest Water Institutions. *Journal of Economic Issues* 20(1): 87–107.

Wescoat, James. 2002. Water Policy and Cultural Exchange: Transferring Lessons from around the World to the Western United States. Paper presented at conference, Allocating and Managing Water for a Sustainable Future: Lessons from around the World, 11–14 June. Natural Resources Law Center, University of Colorado School of Law.

Willey, Zach. 1992. Behind Schedule and Over Budget: The Case of Markets, Water, and the Environment. *Harvard Journal of Law and Public Policy* 15: 391–425.

Williamson, Oliver. 1985. *The Economic Institutions of Capitalism: Firms, Markets, Relational Contracting*. New York: Free Press.

World Bank. 1993. *Water Resources Management: A World Bank Policy Paper*. Washington, DC: World Bank.

———. 1994. *Peru: A User-Based Approach to Water Management and Irrigation Development*. World Bank Report No. 13642-PE. Washington, DC: World Bank.

World Commission on Water for the 21st Century. 2000. *From Vision to Action*. Final Report, Second World Water Forum and Ministerial Conference, 17–22 March 2000, The Hague.

Young, Mike. 1997. *Water Rights: An Ecological Economics Perspective*. Working Paper in Ecological Economics No. 9701. Canberra: Australian National University, Center for Resource and Environmental Studies.

Zegarra, Eduardo. 1997. *Límites y Posibilidades de la Apertura del Mercado de Aguas: Reflexiones sobre la Discusión de una Nueva Ley de Aguas en el Perú 1992–1997*. December. Unpublished document.

INDEX

ABOUT THE AUTHOR

Carl Bauer is a fellow at Resources for the Future. He works in the areas of comparative water law, policy, and management in Latin America and the United States, with a focus on political economy and property rights. Bauer has been a lecturer and post-doctoral researcher in environmental studies at the University of California–Berkeley, as well as a visiting professor and Fulbright Scholar at universities in Chile and Argentina. He has been a consultant for the United Nations, the Global Water Partnership, the World Bank, the Inter-American Development Bank, and other international organizations on matters of water rights, water markets, and economic instruments for water resources management. Bauer's previous book was *Against the Current: Privatization, Water Markets, and the State in Chile*, which has also been published in Spanish. His current research examines the impacts of electricity deregulation on river basin management in South America and tries to bridge the gap between lawyers, economists, and geographers in environmental regulation.